E. Richter · T. Feyerabend

Normal Lymph Node Topography

CT Atlas

Springer

Berlin
Heidelberg
New York
Hong Kong
London
Milan
Paris
Tokyo

E. Richter · T. Feyerabend

Normal Lymph Node Topography

CT Atlas

Foreword by W. Bohndorf

With 66 Figures in 147 Separate Illustrations

 Springer

PROF. DR. MED. ECKART RICHTER
Klinik für Strahlentherapie und Nuklearmedizin
der Universität zu Lübeck
Ratzeburger Allee 160
W-23538 Lübeck
Germany

PROF. DR. MED. THOMAS FEYERABEND
Strahlentherapie Bonn-Rhein-Sieg
Waldstr. 73
W-53177 Bonn
Germany

ISBN 3-540-20857-7 Springer-Verlag Berlin Heidelberg

ISBN 3-540-52549-1 Springer-Verlag Berlin Heidelberg New York Tokyo (First printing)

Library of Congress Cataloging-in-Publication Data
Richter, E. (Eckart), 1940– . Normal lymph node topography: CT-atlas/E. Richter, T. Feyerabend; foreword
by W. Bohndorf. p.; cm. Includes bibliographical references and index. ISBN 3-540-20857-7 (softcover:
alk paper) 1. Lymphatics–Tomography–Atlases. 2. Lymph nodes–Tomography–Atlases. I. Feyerabend, T.
(Thomas), 1957– . II. Title. [DNLM: 1. Lymph Nodes–anatomy & histology–Atlases. 2. Tomography, X-Ray
Computed–Atlases. WH 17 R535n 2004] QM197.R53 2004 611'.42–dc22

Springer-Verlag is a part of Springer Science + Business Media

springeronline.com

© Springer-Verlag Berlin Heidelberg 1991, 2004
Printed in Germany

Cover design: F. Steinen, ᵉStudio Calamar, Spain
Printing and bookbinding: Strauss, Mörlenbach

21/3150 – 5 4 3 2 1 0
Printed on acid-free paper

Preface for the Second Printing

The warm welcome accorded to the original publication of this atlas, coupled with its unavailability despite increasing demand, prompted the publisher to launch this revised reprint. Since 1991, when the book first appeared, radiotherapy has made major advances. Nowadays three-dimensional definition of the target volume based on CT slices has become routine. In addition, new treatment techniques such as stereotactic radiotherapy of the head and body and especially intensity-modulated radiotherapy (IMRT) have evolved. A prerequisite for the clinical application of these techniques is precise definition of the target volume, including the lymphatic pathways.

In the intervening years our knowledge about the topography of the lymphatic drainage of the different organs has not expanded dramatically. Therefore, the text and the figures remain unchanged.

We hope that this reprint once again meets the demand for a comprehensive book on lymphatic anatomy as seen on CT and that the atlas serves as a valuable tool for a new generation of radiotherapists.

E. Richter
T. Feyerabend

Foreword

In oncology the primary tumor and the regional lymphatics constitute a unit, and this fact must be borne in mind in each and every therapeutic approach, especially surgery and irradiation. Oncologists must of course know the exact size, location and histology of the primary tumor, but detailed knowledge of the dissemination of tumor cells in the regional lymph nodes is also indispensable for successful treatment.

Nowadays a variety of methods are available for the diagnosis of tumors and metastases, and radiologic procedures such as computed tomography, magnetic resonance imaging and ultrasound have been very successful in tumor imaging in the last decade. However, the regional lymph vessels and nodes which also have to be dissected or irradiated, owing to the high proportion of cases in which there is microscopic tumor involvement, are still barely accessible to imaging. It is thus extremely important that radiotherapists and surgeons familiarize themselves with the complicated pathways of metastatic dissemination.

The authors, Prof. Richter and Dr. Feyerabend, have worked hard to produce this excellent atlas of lymph node topography. Together, the plates of the topographic anatomy and the CT scans allow ready identification of the different lymph node regions. The comprehensive, systematic approach will be of great value both to students and to experienced clinicians in their daily work.

I am delighted that this volume has its origins in the radiotherapy department of the University of Würzburg, where for many years CT has played an important role in treatment planning. The problem of metastatic lymph node involvement and the individual spatial evaluation of the lymphatics by means of CT scans have been subjects of special interest in Würzburg, and this atlas has its roots in the clinical practice there. I am confident that it will become an invaluable reference for all oncologists.

W. Bohndorf

Preface

Detailed knowledge of lymph node topography is a fundamental prerequisite for the diagnosis and therapy of malignant disease. Anatomic and lymphographic studies of lymph nodes are numerous and have been well documented. However, there has been no presentation of lymph node topography by computed tomography (CT). In this respect this CT atlas is intended to satisfy a long-felt need. Not only does CT represent one of the most common modern imaging techniques, it also bears the crucial advantage of imaging the exact and undistorted anatomy in every single scan. The precise depiction of enlarged lymph nodes facilitates the establishment of tumor stage. Drawings of the selected scans refer to the normal anatomy. Other illustrations depict lymph nodes and groups of lymph nodes in the CT scan according to their topographical location. Thus, the reader will be able to study both lymph node topography and normal anatomy in CT scans.

After an introduction to the lymphatic system the atlas is divided into four chapters covering the head and neck, thorax, abdomen, and pelvis and inguinal region. The lymph nodes of the limbs, which as a rule are clinically less important, are not considered. Each chapter begins with a description of the lymph drainage and the regional lymph nodes for every region and every organ. Additional schematic drawings complement the text. There follow remarks on the topographical anatomy of the individual lymph node groups. Throughout the text, reference is made to the corresponding illustrations in the plates. The plates following the text present on facing pages an axial CT scan, a schematic illustration of this scan (including the names of the important anatomic structures) and another schematic illustration depicting the topography of the individual lymph node groups. Within each chapter, the lymphatic drainage regions and lymph node groups are treated from cranial to caudal.

The reader is recommended to start with the text in each chapter, then to study the topography in the plates, and finally to go through the whole series of plates, thus receiving a three-dimensional impression of the lymph node topography. As for the search for a single piece of information, either the index with the English designations of the lymph nodes or the table of contents with the Latin terms may be consulted. The index refers to the corresponding part of the text and to the appropriate plates.

The Latin names of the lymph nodes follow the latest edition of *Nomina Anatomica* (32). However, some lymph node groups are not mentioned in *Nomina Anatomica* or are subsumed in other terms. The inofficial terms used for these have been set in parentheses in the sections on the topographical anatomy of the lymph node groups, e. g., (Lnn. nuchales). In the lists of the regional lymph node groups (in the sections on lymphatic drainage regions) the lymph node groups in parentheses are those subgroups which are essential for the lymph drainage of this region or organ. For instance, the lymph of the skin of the neck is drained to the Lnn. cervicales superficiales and the Lnn. supraclaviculares, both subgroups of the Lnn. cervicales profundi (see page 9); other groups make hardly any contribution to the lymph drainage of the skin of the neck.

We would like to express our thanks to our secretary, Mrs. Doris Schmidt, for her untiring dedication in typing the manuscript. We are also indebted to Springer-Verlag for kindly consenting to publish the atlas in this generous form. Our special thanks go to Dr. U. Heilmann and Mr. W. Bischoff for their continuous willingness to discuss and help in the preparation and design of the book.

This CT atlas should appeal to every physician who has to deal with the topography of the lymph nodes, especially to radiologists, oncologists, radiotherapists and surgeons. Students will also find it valuable.

A real concern for us has been to make the atlas as useful and informative to as wide a range of readers as possible. Any suggestions and constructive criticism will be highly welcome.

E. Richter
T. Feyerabend

Contents

Introduction 1

The Lymphatic System in General 3

Head and Neck 7

General Considerations 7
Lymphatic Drainage Regions 9
  Scalp 9
  Skin of Neck 9
  Face 9
  Nose 10
  Ear 11
  Orbit 11
  Paranasal Sinuses 11
  Oral Cavity 12
  Tongue 12
  Nasopharynx 13
  Oropharynx 13
  Hypopharynx 13
  Larynx 13
Salivary Glands 15
  Submandibular Gland 15
  Parotid Gland 15
  Thyroid Gland 15
Topographical Anatomy of Regional Lymph Node Groups of Head and Neck . . . 17
  Lnn. Occipitales 17
  Lnn. Mastoidei 17
  Lnn. Parotidei Superficiales et Profundi 17
  Lnn. Faciales 17
  Lnn. Submandibulares 17
  Lnn. Submentales 18
  (Lnn. Sublinguales) 18
  Lnn. Cervicales Anteriores 18

Lnn. Cervicales Laterales 18
(Lnn. Nuchales) 19

Thorax and Breast 41

General Considerations 41
Lymphatic Drainage Regions 41
Lung 41
Mediastinum 44
Heart and Pericardium 46
Thymus Gland 46
Esophagus 46
Costal Wall and Parietal Pleura 48
Diaphragm 48
Breast 48
**Topographical Anatomy of Regional Lymph Node Groups of Mediastinum,
Breast and Axillary Region** 51
Lnn. Cervicales Profundi 51
Lnn. Axillares 51
Lnn. Paramammarii 52
Lnn. Parasternales 52
Lnn. Intercostales 52
Lnn. Mediastinales Anteriores 52
Nodus Ligamentis Arteriosi 52
Lnn. Pericardiales Laterales 52
Lnn. Prepericardiales 52
Lnn. Mediastinales Posteriores 53
Lnn. Prevertebrales 53
Lnn. Phrenici Superiores 53
Lnn. Coeliaci 53

Abdomen 81

General Considerations 81
Lymphatic Drainage Regions 82
Abdominal Wall 82
Liver 82
Gallbladder and Extrahepatic Bile Ducts 84
Stomach 84
Spleen 85
Pancreas 86

Small Bowel 88
Colon 88
Suprarenal Gland 90
Kidney 90
Ureter 91

Topographical Anatomy of Regional Lymph Node Groups of Abdomen 92

Parietal Nodes 92
Lnn. Lumbales Sinistri 92
Lnn. Lumbales Intermedii 92
Lnn. Lumbales Dextri 92
Lnn. Phrenici Inferiores 92
Lnn. Epigastrici Inferiores 93

Visceral Nodes 93
Lnn. Coeliaci 93
Lnn. Hepatici 93
Lnn. Gastrici (Dextri et Sinistri) 93
Lnn. Gastro-omentales (Dextri et Sinistri) 93
Lnn. Pylorici 93
Lnn. Pancreaticoduodenales (Superiores et Inferiores) 94
Lnn. Pancreatici (Superiores et Inferiores) 94
Lnn. Splenici 94
Lnn. Mesenterici 94
Lnn. Appendiculares 94
Lnn. Pre- et Retrocecales 94
Lnn. Ileocolici 94
Lnn. Mesocolici 95
Lnn. Mesenterici Inferiores 95

Pelvis and Inguinal Region 113

General Considerations 113
Lymphatic Drainage Regions 114
Rectum and Anal Canal 114
Urinary Bladder 115
Urethra 117
Male Internal Genitals 117
Prostate 117
Ductus Deferens and Seminal Vesicles 118
Testicle and Epididymis 119
Male External Genitals 120
Skin of Scrotum and Penis 120
Glans and Body of Penis 120

Female Internal Genitals 121
   Ovary 121
   Body of Uterus, Fallopian Tube 122
   Uterine Cervix 123
   Vagina 123
Female External Genitals 124
   Vestibule of Vagina, Labia Pudendi 124
Body and Glans of Clitoris, Bartholin's Glands 124
**Topographical Anatomy of Regional Lymph Node Groups
of Pelvis and Inguinal Region** 125
Parietal Nodes 125
   Lnn. Rectales Superiores 125
   Lnn. Iliaci Communes 125
   Lnn. Iliaci Externi 126
   Lnn. Iliaci Interni 126
Visceral Nodes 127
   Lnn. Paravesiculares 127
   Lnn. Para-uterini 127
   Lnn. Paravaginales 127
   Lnn. Pararectales (Anorectales) 127
Inguinal Nodes 127
   Lnn. Inguinales Superficiales 127
   Lnn. Inguinales Profundi 127

References 149

Subject Index 153

Introduction

The first systematic investigations of the lymphatic system can be credited to Asellius in 1627 [3] and Pecquet in 1653 [33]. In 1787, Mascagni [26] presented a comprehensive essay on the human lymphatic system based upon the visualization of the lymphatics using mercury (Fig. 1). In more recent times, particularly the work of Bartels [4], Rouvière [42] and Haagensen [15] deserves mention. In 1932, Shdanov [45] was one of the first to succeed in visualizing parts of the lymphatic system roentgenologically, employing compounds of mercury, lead or silver. In 1952, it was Kinmonth [19] who introduced the routine use of lymphography. In the following years the method steadily became more sophisticated, especially with the introduction of oily contrast media. The visualization of the lymphatic drainage system then began to attract great clinical interest because of its potential in the diagnosis and therapy of malignancies. The considerable time taken up by the examination and the occasional failures due to imperfect technique limit the usefulness of lymphography. Another disadvantage is the fact that with bipedal lymphography only the inguinal, iliac and retroperitoneal lymph nodes up to the cisterna chyli can be demonstrated, as from the cisterna chyli the lymph flows into the thoracic duct without passing through any more lymph nodes. There were many attempts to depict lymph nodes in other regions by lymphography. By injecting oily contrast media into a retroauricular lymph vessel it is possible to visualize groups of cervical lymph nodes, but this diagnostic method is difficult and failed to become popular. The same applies to lymphography of the upper limb and to direct lymphography of the female breast [18].

Interstitial lymphoscintigraphy enables detection of a lack of tracer uptake or displacements of lymph nodes. The differentiation of single lymph nodes is not possible, as the spots of activity are always due to groups of nodes. Up to now no routine, reproducible scintigraphic procedure exists which shows groups of lymph nodes in their actual location in the patient. All images project the findings onto the surface of the patient's body; topographic studies are not possible.

The sonography of lymph nodes has become a reliable method of clinical diagnostic investigation, especially for the abdominal, retroperitoneal and cervical lymph nodes. Although the lower limit of spatial resolution is on the order of 2 mm, and although the topographic assignment of vessels, parenchymal organs and enlarged lymph nodes is relatively easy, this method also fails to visualize the normal topography of the lymphatic system in a reproducible manner.

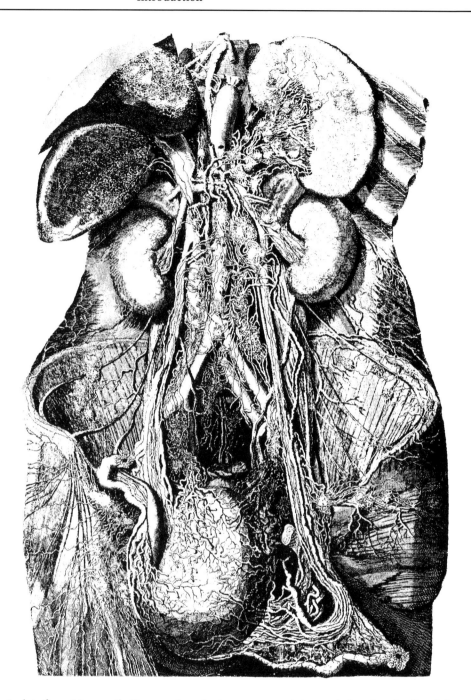

Fig. 1. A plate from Mascagni's *Vasorum lymphaticorum corporis humani*, showing details of the collecting lymph channels of the internal genitals [26]

The introduction of computed tomography (CT) in 1974 raised great hopes. This method permits transverse tomograms of the whole body with increments of 1.5–12 mm. In practice, structures as small as 3 mm in diameter can be discriminated, thus nourishing the hope that CT might also visualize small, pathologically unchanged lymph nodes. Widespread experience, however, tempered this initial enthusiasm, which gave way to a more realistic assessment of the capabilities of this method. Lymph nodes more than 10 mm in diameter can be reliably depicted in nearly all anatomic regions, but lymph nodes up to 5 mm are inconstantly visualized. Although CT fails to record the lymphatic system completely, the lymph nodes that are imaged can be unequivocally related to the surrounding anatomic structures. The latter is achieved because, in contrast to conventional tomography, CT allows the visualization of the anatomical structures in the chosen slice position without superimposition. The topography of lymph nodes can be fully demonstrated by means of the complete image series of the normal anatomy, even if single lymph nodes are not visible on the CT scans.

The Lymphatic System in General

Today it is generally accepted that the lymphatic drainage system originates from lymph capillaries, which begin as blind pouches and are coated by endothelium. These capillaries make up a coarse-meshed network and join to form lymph vessels. By way of these vessels the lymph reaches the lymph nodes. These nodes are called primary regional lymph nodes if they take up the lymph of a circumscribed capillary network (tributary region). Figure 2 shows the basic principle of organization of the lymphatic drainage system, as described by Kubik et al. [22, 24]. Each region of the lymphatic drainage system has its own primary lymph nodes, which the lymph then passes to secondary and tertiary lymph nodes. There are connections between the different lymphatic regions at the level of the primary lymph nodes as well as between subsequent lymph vessels. For example, the lymph node 3 b receives lymph from six drainage regions, but the tributary region of this node is solely represented by the capillary network F.

From the secondary and tertiary and to a lesser extent from the primary lymph nodes the lymph is mainly delivered to the venous blood by way of four pairs and one group of main trunks (trunci lymphatici) and by two main ducts (ductus lymphatici) (Fig. 3):

- The jugular trunks (trunci jugulares) each receive lymph from 45–85 lymph nodes and drain into the venous angle.
- The subclavian trunks (trunci subclavii) each receive lymph from 20–50 lymph nodes and drain mainly into the lymphatic duct on the right and into the venous angle on the left.
- The lumbar trunks (trunci lumbales) each receive lymph from 70–110 lymph nodes and drain into the cisterna chyli.

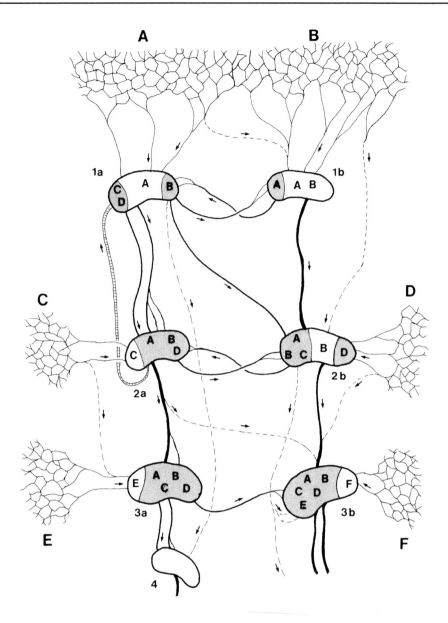

Fig. 2. Basic organization of the lymphatic drainage system. See text for explanation. (Modified from [22])

- The group of intestinal trunks (trunci intestinales) receive lymph from 150–300 lymph nodes and empty into the cisterna chyli.
- The bronchomediastinal trunks (trunci bronchomediastinales), each receive lymph from 30–55 lymph nodes and empty into the right lymphatic duct on the right and the thoracic duct on the left.
- The right lymphatic duct (ductus lymphaticus dexter) drains into the right venous angle.
- The thoracic duct (ductus thoracicus) receives lymph from 20-100 lymph nodes and terminates at the left venous angle.

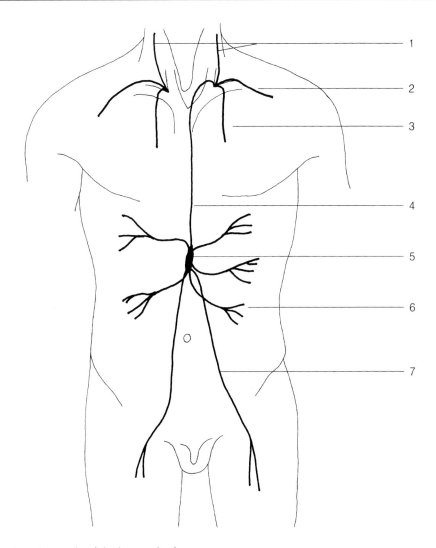

Fig. 3. The main lymph vessels of the human body

1 Jugular trunk
2 Subclavian trunk
3 Bronchomediastinal trunk
4 Thoracic duct

5 Cisterna chyli
6 Intestinal trunks
7 Lumbar trunk

Figure 4 shows schematically the draining of the trunks and ducts into the venous angle (the junction of the internal jugular vein and the subclavian vein), neglecting numerous variants.

In accordance with their centripetal flow the lymphatic pathways can be divided into two main sections: one of the upper and one of the lower half of the body. The upper part includes the head, the neck, the upper limbs, the upper part of the trunk and the thoracic cavity. The lower part comprises the lower limbs, the lower part of the trunk, the abdominal cavity and the pelvis.

Lymph from the head and neck drains into the jugular trunk and the subclavian trunk by two routes: medially via the lymph nodes along the internal jugular vein and

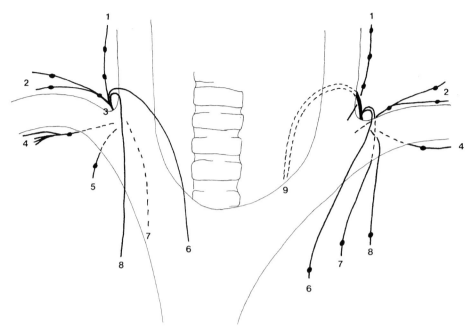

Fig. 4. The termination of the main lymph vessels of the neck at the venous angle. (Modified from [15])

1 Jugular trunk
2 Transverse cervical trunk
3 Right lymphatic duct
4 Subclavian trunk
5 Intercostal trunk

6 Anterior mediastinal trunk
7 Tracheobronchial trunk
8 Internal mammary trunk
9 Thoracic duct

laterally via the chain of lymph nodes in close proximity to the accessory nerve and the subclavian vein. The lymph of the upper limb drains into the venous angle via the jugular trunk. From the thoracic cavity the lymph of the heart, mediastinum and lungs is collected by the bronchomediastinal trunk, while the lymph of the intercostal regions empties in the upper part into the intercostal trunk or directly into the venous angle and in the lower part into the thoracic duct. The lymph of the parasternal region on the left side is drained into the thoracic duct and on the right side into the right lymphatic duct. In the upper part of the trunk the lymph drainage of the female breast by the subclavian and the parasternal route is of special importance.

From the lower limbs, the pelvic viscera and the urogenital system the lymph is transported by way of the lumbar trunk to the cisterna chyli. The intestinal trunks, taking up the lymph of the small and large intestine, also open into the cisterna chyli. From the cisterna chyli, which is the enlarged origin of the thoracic duct, the lymph is conveyed via the thoracic duct to the left venous angle.

Due to the numerous anatomical variations the lymphatic system cannot be forced into a rigid framework. Therefore Kubik [20] emphasized that, although the connections of a single regional lymph node with a definite region or a definite organ are relatively constant, these correlations must not be seen in a stereotyped manner.

Head and Neck

General Considerations

The head and neck area is rich in lymph nodes and lymph vessels. This region accounts for 30 % of all lymph nodes in man. In comparison to other regions the lymph nodes in the head and neck are small. They vary from 2 to 15 mm in diameter, and are thus also small in volume, in 75 % of cases less than 0.5 cm^3.

In general, the lymphatic flow of the head and neck is craniocaudal in direction: the one exception is that at the border of the neck and the trunk the flow is transverse. In contrast to the numerous variations at other sites, there are clear topographical relations between the lymph nodes and the internal jugular vein, the accessory nerve, the digastric muscle and the omohyoid muscle (Fig. 5).

Rouvière's classification of the head and neck lymph nodes first published in 1932 [42] and separating 10 groups is still valid today. However, Rouvière did not describe the group of the Lnn. nuchales, whose existence has meanwhile been proved and which can be dissected. Depending on the anatomy a horizontal chain of lymph nodes may be found at the border of the head and the neck. The craniocaudal transport of the lymph to the base of the neck is accomplished via the main lymph draining paths which are associated with the lateral cervical lymph nodes: vertically (along the internal jugular vein), obliquely (along the accessory nerve) and transversely. On each side of the neck the main lymph vessels run to the jugulosubclavian junction and empty there as the jugular trunk into the veins (Figs. 4, 5). The efferent vessels of the transverse cervical chain meet the vertical chain at the junction of the internal jugular vein and the facial vein. The lymph nodes of this region, i.e. the cranialmost of the deep cervical lymph nodes, the so-called jugulodigastric lymph nodes, are sometimes classified as lymph nodes of the junction [12]. Clinically, these nodes are extremely important, as they drain the lymph of nearly all parts of the head and neck to the more caudal regions. According to Fisch and Sigl [12], these are the regional lymph nodes for the nasopharynx, the tonsils, the maxillary sinus, the hard and soft palate, the base of the tongue, the mesopharynx, the larynx and the vestibule of the larynx. Due to their cranial position it is evident that these lymph nodes may be the starting-point for tumor cells spreading into the main lymph pathways [22].

The individual lymph node groups have constant topographical relations to the surrounding tissue, so they are easily localized by reference to anatomical landmarks such as blood vessels, muscles or glands. It is largely owing to Rouvière [42] and Taillens [47] that the topographical anatomy and functional relationships have been analyzed systematically.

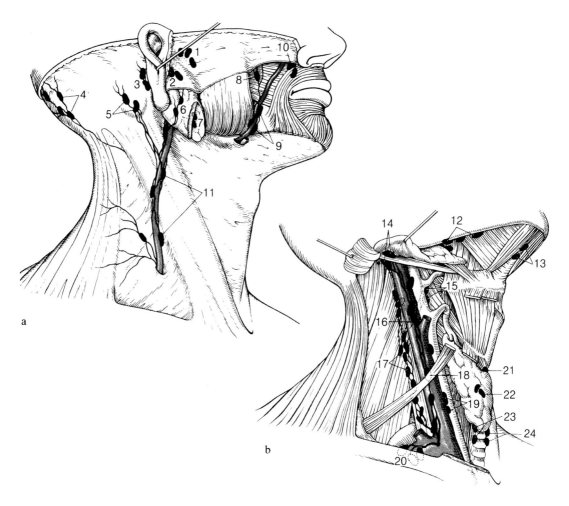

Fig. 5a,b. The superficial (a) and deep (b) lymph nodes in the head and neck.
(Adapted from [11])

 1 Pre-auricular lymph nodes
 2 Superficial parotid lymph nodes
 3 Infra-auricular lymph nodes
 4 Occipital lymph nodes
 5 Retro-auricular (mastoidal) lymph nodes
 6 Deep parotid lymph nodes
 7 Intraglandular parotid lymph nodes
 8 Buccinator lymph node
 9 Mandibular lymph node
10 Nasolabial lymph node
11 Superficial lymph nodes
12 Submandibular lymph nodes

13 Submental lymph nodes
14 Jugulodigastric lymph nodes
15 Retropharyngeal lymph nodes
16 Superficial cervical lymph nodes
17 Lateral jugular lymph nodes
18 Juguloomohyoid lymph node
19 Anterior jugular lymph nodes
20 Prelaryngeal lymph node
21 Supraclavicular lymph nodes
22 Thyroid lymph nodes
23 Paratracheal lymph nodes
24 Pretracheal lymph nodes

Lymphatic Drainage Regions

Scalp

The lymph vessels of the forehead and the glabella run laterally to the preauricular lymph nodes. The parietal part of the scalp is drained by lymph vessels which empty into the postauricular nodes (Lnn. mastoidei) or the deep cervical lymph nodes. The collecting vessels of the occipital zone drain to the occipital lymph nodes.

Regional Lymph Nodes *Lnn. cervicales profundi*
            Lnn. mastoidei

Skin of Neck

The lymph vessels of the nape and the side of the neck run to the superficial cervical lymph nodes. The supraclavicular region, the skin over the sternocleidomastoid muscle, the suprahyoid and the infrahyoid region are drained along the internal jugular vein to the supraclavicular lymph nodes.

Regional lymph nodes *Lnn. cervicales superficiales*
           Lnn. cervicales profundi
           (Lnn. supraclaviculares)

Face

The lateral collecting lymph vessels of the upper and lower eyelid and the conjunctiva drain to the superficial parotid lymph nodes which are found at pre- and post-auricular locations. The medial collecting trunks of the eyelids run to the submandibular lymph nodes. The lymph of the lacrimal gland is drained to the pre-auricular and submandibular lymph nodes, which also take the lymph vessels of the skin of the cheek and the buccal mucosa, excepting the skin of the anterior parts of the cheeks, whose lymph vessels empty into the submental lymph nodes. The chief lymphatic pathways of the lower lip, the mucosa of the lower lip and the chin go to the submental lymph nodes and the submandibular lymph nodes; at the midline many collecting vessels cross to the contralateral side. The pathway of lymph from the upper lip and its mucosa is predominantly to the submandibular lymph nodes and rarely to the submental lymph nodes and the infra-auricular deep parotid lymph nodes.

Regional lymph nodes *Lnn. parotidei superficiales*
           Lnn. parotidei profundi
           Lnn. submandibulares
           Lnn. submentales

Nose

The lymph of the skin of the nose and the anterior parts of the mucosa of the nose runs to the ventral and dorsal lymph nodes of the submandibular gland. The medial and posterior parts of the mucosa are drained by the cranial lymph nodes of the internal jugular vein and the retropharyngeal lymph nodes (Fig. 6).

Regional lymph nodes   *Lnn. submandibulares*
                           Lnn. cervicales profundi craniales
                           Lnn. retropharyngeales

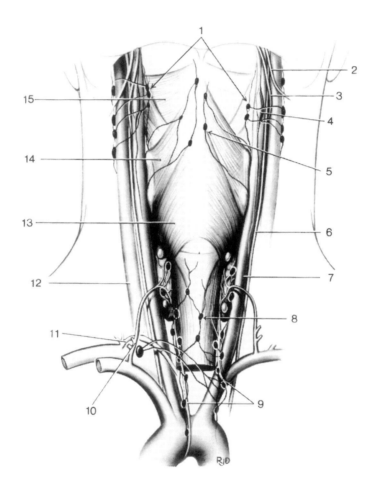

Fig. 6. Topographical anatomy of the retropharyngeal lymph nodes: posterior view after removing the cervical vertebrae and the paraspinal musculature. (Adapted from [15])

1 Lateral retropharyngeal nodes
2 Spinal access. n.
3 Hypoglossal nerve
4 Superior cervical ganglion
5 Medial retrophar. nodules
6 Vagus nerve
7 Rt. carotid artery
8 Inferior medial retropharyngeal and esophageal nodes

9 Recurrent laryngeal nerves and nodes
10 Inferior thyroid artery
11 Thoracic duct
12 Jugular vein
13 Inferior pharyngeal const. muscle
14 Middle pharyngeal const. muscle
15 Superior pharyngeal const. muscle

Ear

The anterior parts of the ear are, drained to the pre-auricular, the lower parts to the infra-auricular and the posterior parts to the retro-auricular lymph nodes (Lnn. mastoidei). The lymph vessels of the external auditory canal run to pre- and infra-auricular lymph nodes, intraglandular lymph nodes and the deep lymph nodes of the internal jugular vein (Lnn. cervicales profundi). The collecting vessels of the eustachian tube go directly to the retropharyngeal nodes.

Regional lymph nodes   *Lnn. mastoidei*
                             Lnn. parotidei superficiales et profundi
                             Lnn. cervicales profundi craniales
                             Lnn. retropharyngeales

Orbit

The cornea, the sclera, the lens and the retina do not contain true lymphatic capillary vessels. The conjunctiva has rich lymphatics which extend circumferentially around the cornea (circulus lymphaticus). Collecting lymph vessels run to the inner and outer canthus and then the lymph is drained to the submandibular and parotid lymph nodes respectively.

Regional lymph nodes   *Lnn. submanibulares*
                             Lnn. parotidei superficiales et profundi

Paranasal Sinuses

The lymph drains to the connections between the paranasal sinuses and their nasal openings and is collected in the lymph nodes of the internal jugular vein and the retropharyngeal nodes.

Regional lymph nodes   *Lnn. cervicales profundi craniales et caudales*
                             Lnn. retropharyngeales

Oral Cavity

The lymph of the buccal mucosa is drained to the submandibular lymph nodes, that of the outer parts of the alveolar ridge either to the preglandular (anterior part of mucosa) or to the retroglandular (posterior part of mucosa) submandibular lymph nodes. The lymph vessels of the inner alveolar ridge and the hard and soft palate run to the submandibular lymph nodes, the retropharyngeal nodes and those along the internal jugular vein. At the midline the vessels cross over. The lymph from the gingiva of the mandible goes to the submandibular lymph nodes and the nodes of the internal jugular vein. The lymph of the floor of the mouth is collected by sublingual nodes, the preglandular submandibular lymph nodes and those of the internal jugular vein; a rare pathway is the one to the submental nodes.

Regional lymph nodes   *Lnn. submandibulares*
                                        Lnn. submentales
                                        Lnn. sublinguales
                                        Lnn. cervicales profundi craniales
                                        Lnn. retropharyngeales

Tongue

The lymphatic pathways of the tongue vary considerably. The lymph vessels of the tip of the tongue drain to the submental lymph nodes and the caudal nodes of the internal jugular vein. The lymph of the lateral parts of the tongue chiefly goes to the submandibular lymph nodes and the cranial nodes of the internal jugular vein. The central parts of the tongue drain to the lymph nodes of the internal jugular vein and the posterior third of the tongue to the cranial lymph nodes of the internal jugular vein. At the midline the lymph vessels of the central parts of the tongue cross over to the contralateral side.

Regional lymph nodes   *Lnn. submentales*
                                         Lnn. submandibulares
                                         Lnn. cervicales profundi craniales et caudales

Nasopharynx

The lateral and the posterior wall of the pharynx and the region of the eustachian tube drain along the retropharyngeal lymph nodes to the deep cervical nodes. In the midline there are rich connections between the two sides of the nasopharynx.

Regional lymph nodes   *Lnn. retropharyngeales*
                            Lnn. cervicales profundi craniales

Oropharynx

The lymph vessels penetrate the wall of the pharynx and mainly travel along to the jugu-lodigastric lymph nodes which are part of the deep cervical lymph nodes and can be found posterior to the angle of the mandible.

Regional lymph nodes   *Lnn. submandibulares*
                            Lnn. cervicales profundi
                            (Lnn. jugulodigastrici)

Hypopharynx

The lymph of the piriform fossae and the adjacent tissue is collected to lymph nodes located below the hyoid. The paratracheal lymph nodes account for the drainage of the whole hypopharynx.

Regional lymph nodes   *Lnn. cervicales anteriores*
                            Lnn. cervicales profundi
                            Lnn. paratracheales
                            Lnn. retropharyngeales

Larynx (Fig. 7)

In the larynx, different lymphatic pathways have to be considered [28, 29, 41], the border between them being represented by the vocal fold. However, anastomoses between the supraglottic and the infraglottic region have been described [7]. In the supraglottic region the lymph vessels drain to the deep cervical lymph nodes and in the epipharynx and the aryepiglottic folds to the anterior cervical lymph nodes. Inferior to the vocal fold the prelaryngeal lymph nodes are the first draining nodes, the next are the pretracheal nodes and then follow the deep cervical nodes. In the posterior parts of the larynx the lymph is collected by the paratracheal nodes and then passes to the supraclavicular

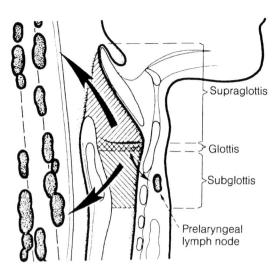

Fig. 7. Lymph drainage of the endolarynx to the deep cervical lymph nodes (large arrows) and the prelaryngeal node. (Adapted from [30])

nodes. Recent studies by Mann [25] show that in the larynx the midline is no barrier; drainage to the contralateral side is possible. There is no crossover between supraglottic and glottic region.

Regional lymph nodes   *Lnn. cervicales profundi*
(Lnn. supraclaviculares)
Lnn. cervicales anteriores
(Lnn. infrahyoidales
Lnn. prelaryngeales
Lnn. pretracheales
Lnn. paratracheales)

Salivary Glands

Submandibular Gland

The preglandular lymph nodes collect the lymph of the upper and lateral part of this gland whereas the posterior part is drained to the subdigastric lymph nodes of the internal jugular vein.

Regional lymph nodes   *Lnn. submandibulares*
                        Lnn. cervicales profundi craniales

Parotid Gland

The lymph is drained to the intra- and paraglandular lymph nodes and then passes to the deep internal jugular chain.

Regional lymph nodes   *Lnn. parotidei superficiales et profundi*
                        Lnn. cervicales profundi craniales

Thyroid Gland (Fig. 8)

The lymphatic capillaries in the thyroid gland form a very rich network. The lateral superior parts of the gland drain to the lymph nodes of the internal jugular vein, whereas the lymph vessels of the medial superior regions run to the prelaryngeal nodes. As for the medial inferior regions, the collecting lymph vessels run both to the para- and pretracheal nodes and to the intraglandular nodes. In the lateral inferior regions the lymph is collected to the inferior nodes of the internal jugular vein.

Regional lymph nodes   *Lnn. cervicales anteriores*
                        (Lnn. prelaryngeales
                        Lnn. pre- et paratracheales
                        Lnn. thyroidei)
                        Lnn. cervicales profundi craniales et caudales

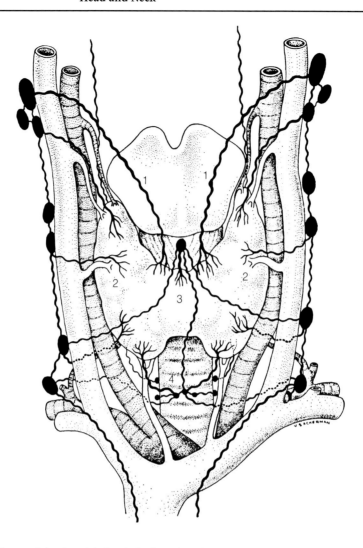

Fig. 8. Lymph drainage of the thyroid gland: the lymph is directed either to the deep internal jugular chain (1, 2) or to the prelaryngeal, pretracheal and paratracheal lymph nodes (3, 4). (Adapted from [1])

Topographical Anatomy of Regional Lymph Node Groups of Head and Neck (Figs. 9–18)

Ten main groups of lymph nodes can be differentiated, some of which can be subdivided. Figure 5 illustrates the topographical anatomy of the main lymph node groups. Their locations and their relations to the surrounding tissue are as follows.

Lnn. Occipitales (Figs. 10–12)

Superficial group: between the posterior insertion of the sternocleidomastoid muscle, the anterior insertion of the trapezius muscle and the occipital aspect of the occipitofrontal muscle.

Deep group: located beneath the insertion of the splenius capitis muscle.

Lnn. Mastoidei (Figs. 10–12)

These are found behind the auricle, just below the posterior auricular muscle.

Lnn. Parotidei Superficiales et Profundi (Figs. 11–13)

There are four subgroups of these nodes
- Subcutaneous, extrafascial and pre-auricular (Fig. 10)
- Subfascial (under the fascia parotidea) and pre-auricular
- Intraglandular (Figs. 11, 13)
- Infra-auricular (Figs. 12, 13), adjacent to the point where the external jugular vein exits from the parotid gland

Lnn. Faciales

This inconstant group of lymph nodes is situated between the skin of the face and the mimetic muscles. They can be found adjacent to the mandible (Ln. mandibularis, Fig. 12), superficial to the buccinator muscle (Ln. buccinatorius, Fig. 11), in the nasolabial groove (Ln. nasolabialis) and, rarely, inferior to the outer canthus (Ln. malaris).

Lnn. Submandibulares (Figs. 12–15)

These nodes may be found in pre- and postvascular (anterior fascial vein), in pre- and retroglandular and also in intracapsular locations (Lnn. paramandibulares, Figs. 13–15).

Lnn. Submentales (Figs.13–15)

These lie between the mylohyoid muscles and both bellies of the anterior digastric muscle, covered by the platysma.

(Lnn. Sublinguales) (Fig. 12)

The location of this group of nodes is inconstant; mainly they can be found at the inferior and lateral borders of the tongue.

Lnn. Cervicales Anteriores

This group, also known as the medial lymph node group, can be divided into several subgroups.

Lnn. superficiales (Figs. 15, 16): along the anterior jugular vein (anterior jugular chain).

Lnn. suprasternales (inconstant group, Fig. 18): in the suprasternal space between the sternum and the sternocleidomastoid muscles.

(Lnn. infrahyoidales): inferior to the hyoid bone and beneath the infrahyoid muscles.

Lnn. prelaryngeales (Fig. 16): over the cricothyroid ligament between the inferior part of the thyroid cartilage and the upper part of the cricoid cartilage.

Lnn. pretracheales (Figs. 17, 18): in front of and alongside the trachea.

Lnn. paratracheales (recurrent nerve chain group, Figs. 17, 18, 26): lateral to the trachea and in close relation to the recurrent nerve.

Lnn. thyroidei (Fig. 18): along the thyroid gland.

Lnn. Cervicales Laterales

This group can be divided into two main groups.

Lnn. cervicales superficiales (Figs. 14, 15): along the external jugular vein.

Lnn. cervicales profundi: these nodes receive lymph from almost all head and neck nodes. They can be divided into several subgroups. One of these, the so-called spinal accessory chain (Figs. 13–18), can be found below the insertion of the sternocleidomastoid muscle and accompanying the accessory nerve posteriorly beneath the trapezius muscle. The other subgroups are as follows:

Lnn. jugulares anteriores et laterales craniales (Figs. 12–14): grouped around the internal jugular vein extending from the base of the skull to the omohyoid muscle.

Lnn. jugulares anteriores et laterales caudales (Figs. 15–18): also along the internal jugular vein, but extending inferior to the omohyoid muscle.

Lnn. supraclaviculares (transverse cervical chain, Figs. 17, 18): superficial to and along the transverse cervical artery and vein and extend to the jugulosubclavian junction.

Lnn. jugulodigastrici: the most cranial of the deep cervical lymph nodes, situated on the internal jugular and internal carotid vessels where the posterior belly of the digastric muscle crosses these vessels.

Ln. jugulo-omohyoideus: between the omohyoid muscle and the internal jugular vein.

Lnn. retropharyngeales (Figs. 6, 11–18): behind the pharynx, in front of the prevertebral fascia and muscles (Lnn. retropharyngeales mediales, Figs. 11–17) and along the internal carotid artery from the base of the skull down to the thoracic inlet (Lnn. retropharyngeales laterales, Figs. 11–15).

(Lnn. Nuchales) (Figs. 13-16)

These nodes may be found beneath the trapezius muscle.

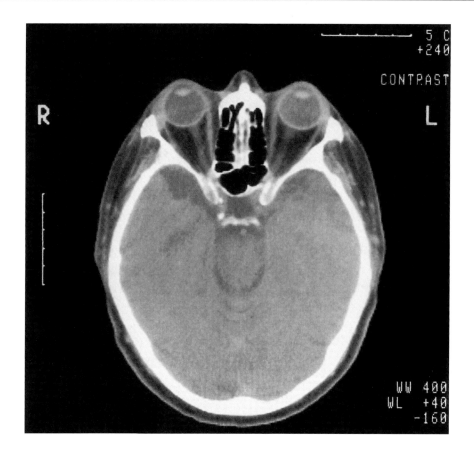

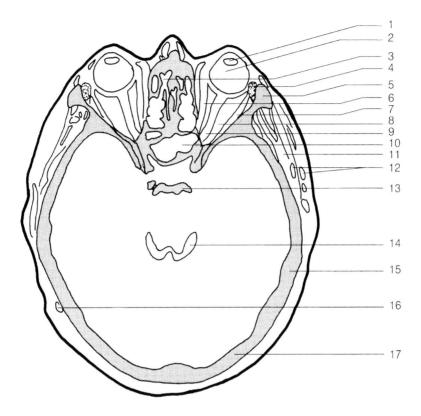

Fig. 9. Axial section of the head

  1 Lens
  2 Vitreous chamber
  3 Ethmoidal cells
  4 Lacrimal gland
  5 Zygomatic bone
  6 Medial rectus muscle
  7 Lateral rectus muscle
  8 Fatty body of orbit
  9 Optical nerve
10 Sphenoidal sinus
11 Temporal muscle
12 Temporal artery and vein
13 Dorsum sellae
14 Fourth ventricle
15 Parietal bone
16 Occipital artery (occipital branch)
17 Occipital bone

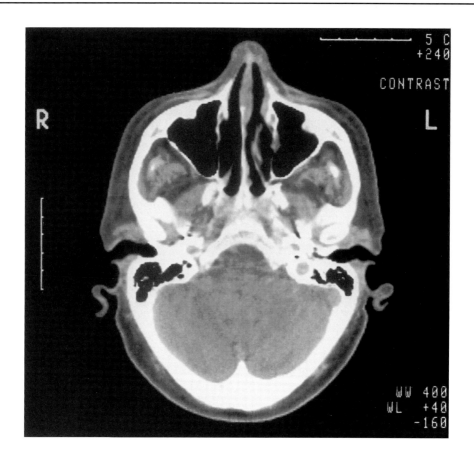

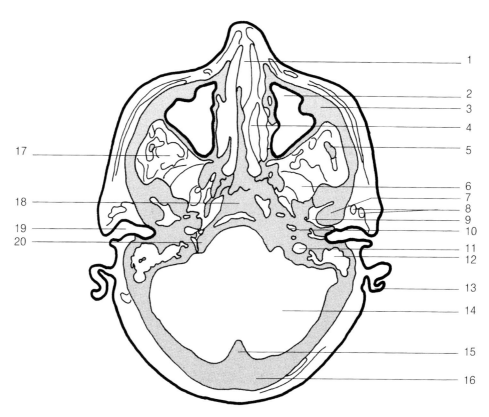

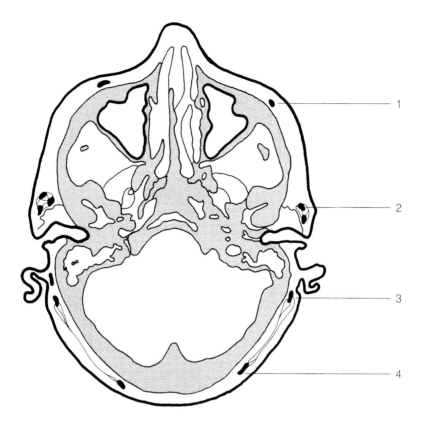

Fig. 10. Axial section of the head ▲

◀ 1 Nasal septum
   2 Maxillary sinus
   4 Medial nasal concha
   3 Zygomatic bone
   5 Coronoid process
   6 Lateral pterygoid muscle
   7 Head of mandible
   8 Temporal artery and vein
   9 Mandibular joint
  10 Internal carotid artery
  11 Upper bulb of jugular vein
  12 Mastoid air cells
  13 Helix
  14 Cerebellum
  15 Internal occipital protuberance
  16 Occipital bone
  17 Temporal muscle
  18 Sphenoidal bone
  19 External auditory meatus
  20 Basilar part of occipital bone

1 Facial lymph nodes
2 Pre-auricular parotid lymph nodes
3 Mastoidal lymph nodes
4 Occipital lymph nodes

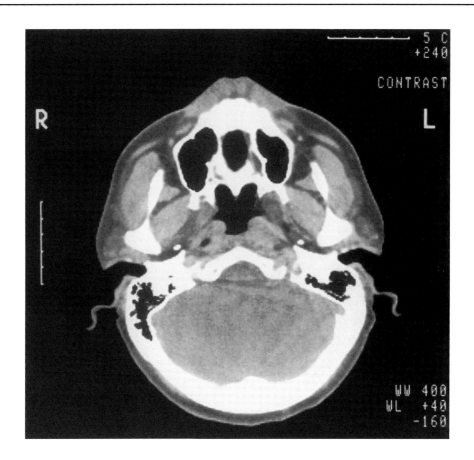

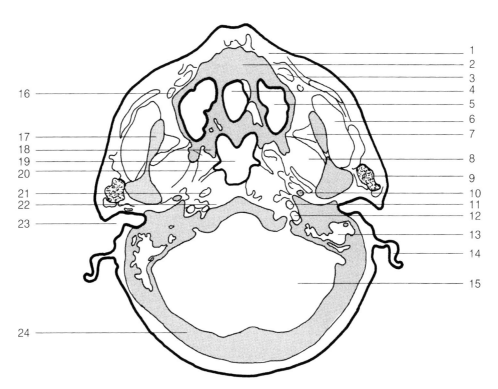

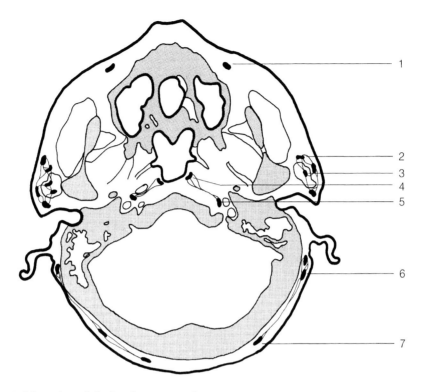

Fig. 11. Axial section of the head ▲

◄ 1 Nasal muscle
  2 Maxilla
  3 Levator muscle of angle of mouth
  4 Oral cavity proper
  5 Maxillary sinus
  6 Masseter muscle
  7 Temporal muscle
  8 Lateral pterygoid muscle
  9 Parotid gland
 10 Styloid process of temporal bone
 11 Internal carotid artery
 12 Internal jugular vein
 13 Mastoid air cells
 14 Helix
 15 Cerebellum
 16 Facial artery and vein
 17 Ramus of mandible
 18 Medial and lateral pterygoid process
 19 Pharynx
 20 Medial pterygoid muscle
    and tensor veli palatini muscle
 21 Recess of pharynx
 22 Long muscle of head
 23 External auditory meatus
 24 Occipital bone

 1 Buccinator lymph node
 2 Superficial and deep pre-auricular
   parotid lymph nodes
 3 Deep intraglandular parotid lymph nodes
 4 Medial retropharyngeal lymph nodes
 5 Lateral retropharyngeal lymph nodes
 6 Mastoidal lymph nodes
 7 Occipital lymph nodes

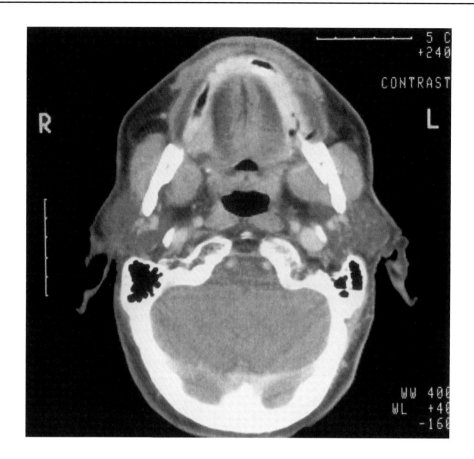

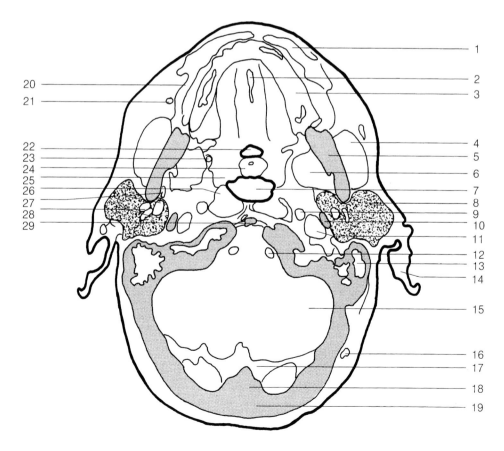

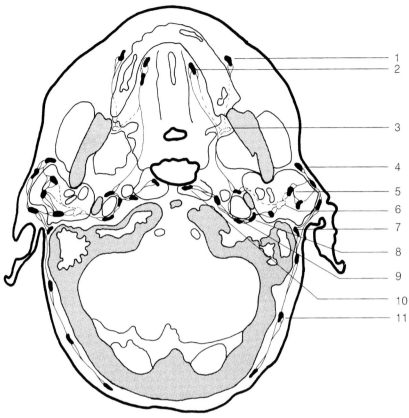

Fig. 12. Axial section of the head ▲

◀ 1 Depressor muscle of lower lip
2 Septum of tongue
3 Alveolar process of mandible
4 Masseter muscle
5 Ramus of mandible
6 Medial pterygoid muscle
7 Pharynx
8 Parotid gland
9 Internal carotid artery
10 Styloid process of temporal bone
11 Internal jugular vein
12 Vertebral artery
13 Mastoid air cells
14 Helix
15 Cerebellum
16 Occipital artery
17 Tentorium cerebelli
18 Internal occipital protuberance
19 Occipital bone
20 Buccinator muscle
21 Facial artery
22 Oral cavity
23 Pterygoid hamulus
24 Uvula
25 Superior constrictor muscle of pharynx
26 Long muscle of head
27 Temporal artery and vein
28 Maxillary artery and vein
29 Odontoid process of axis

1 Mandibular lymph node
2 Sublingual lymph nodes
3 Submandibular lymph nodes
4 Superficial and deep pre-auricular
   parotid lymph nodes
5 Deep intraglandular parotid lymph nodes
6 Deep infra-auricular parotid lymph nodes
7 Mastoidal lymph nodes
8 Anterior and lateral superior jugular
   lymph nodes
9 Lateral retropharygeal lymph nodes
10 Medial retropharyngeal lymph nodes
11 Occipital lymph nodes

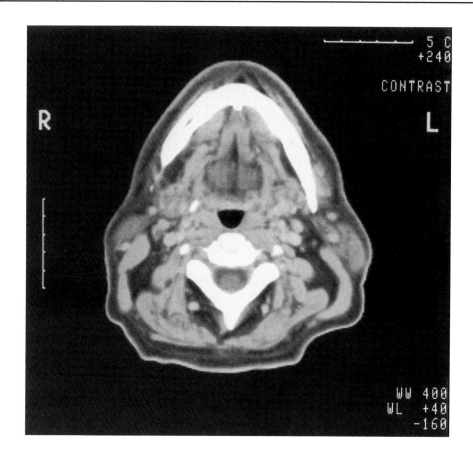

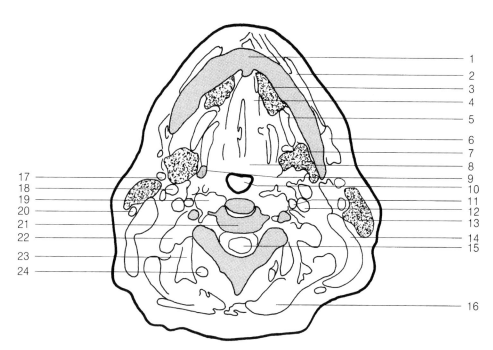

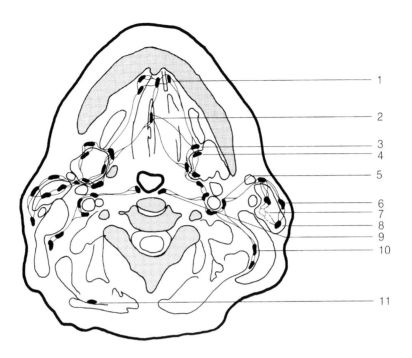

Fig. 13. Axial section of the head ▲

◄ 1 Mandible
   2 Depressor muscle of lower lip
   3 Sublingual gland
   4 Genioglossus muscle
   5 Mylohyoid muscle
   6 Masseter muscle
   7 Submandibular gland
   8 Root of tongue
   9 Styloid process of temporal bone
  10 Pharynx
  11 Internal carotid artery
  12 Internal jugular vein
  13 Parotid gland
  14 Sternocleidomastoid muscle
  15 Cervical spinal cord
  16 Splenius capitis muscle
  17 Retromandibular vein
  18 External carotid artery
  19 Long muscle of head, long muscle of neck
  20 Occipital vein
  21 Cervical vertebra
  22 Levator muscle of scapula
  23 Semispinalis capitis muscle
  24 Deep cervical vein

1, 2 Submental lymph nodes
   3 Submandibular lymph nodes
   4 Paramandibular lymph nodes
   5 Anterior and lateral superior jugular
      lymph nodes
   6 Superficial and deep pre-auricular
      parotid lymph nodes
   7 Deep infra-auricular lymph nodes
   8 Deep intraglandular parotid lymph nodes
   9 Medial retropharyngeal lymph nodes
  10 Deep cervical lymph nodes
  11 Nuchal lymph nodes

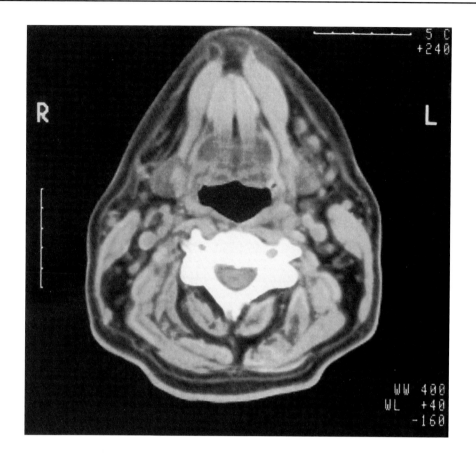

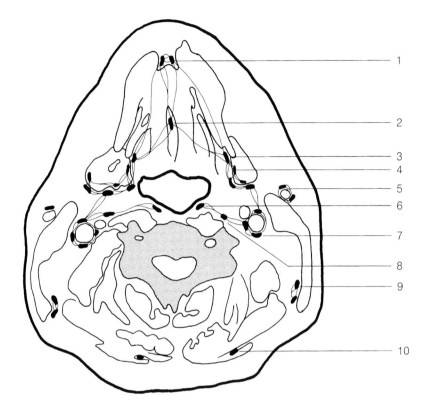

Fig. 14. Axial section of the neck ▲

◄ 1 Anterior belly of digastric muscle
   2 Platysma
   3 Geniohyoid muscle
   4 Anterior jugular vein
   5 Facial vein
   6 Hyoglossus muscle
   7 Retromandibular vein
   8 Submandibular gland
   9 Facial artery
  10 External jugular vein
  11 Internal carotid artery
  12 Internal jugular vein
  13 Sternocleidomastoid muscle
  14 Third cervical vertebra
  15 Levator muscle of scapula
  16 Transversospinalis muscle
  17 Trapezius muscle
  18 Pharynx
  19 Lingual artery
  20 External carotid artery
  21 Long muscle of head, long muscle of neck
  22 Vertebral artery
  23 Vertebral vein
  24 Levator muscle of scapula
  25 Spine of a vertebra

1, 2 Submental lymph nodes
   3 Submandibular lymph nodes
   4 Paramandibular lymph nodes
   5 Superficial lateral cervical lymph nodes
   6 Medial retropharyngeal lymph nodes
   7 Superior jugular lymph nodes
   8 Lateral retropharyngeal lymph nodes
   9 Deep cervical lymph nodes
  10 Nuchal lymph nodes

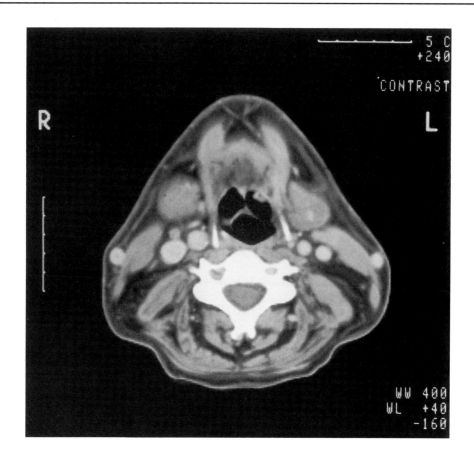

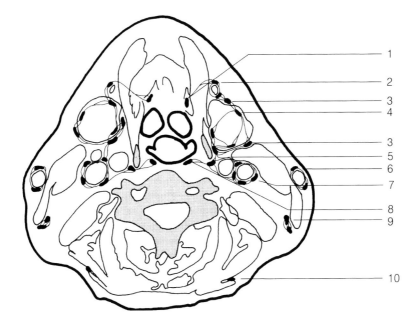

Fig. 15. Axial section of the neck ▲

◄ 1 Anterior belly of digastric muscle
2 Geniohyoid muscle
3 Anterior jugular vein
4 Platysma
5 Glossoepiglottic fold
6 Vallecula epiglottica
7 Submandibular gland
8 Greater horn of hyoid bone
9 Sternocleidomastoid muscle
10 Internal carotid artery
11 Internal jugular vein
12 External jugular vein
13 Long muscle of head,
   long muscle of neck
14 Vertebral canal
15 Levator muscle of scapula
16 Deep cervical vein
17 Multifidus muscle
18 Splenius capitis muscle
19 Trapezius muscle
20 Nuchal ligament
21 Epiglottis
22 Pharynx
23 Superior thyroid artery
24 Vertebral artery
25 Cervical vertebra
26 Middle scalene muscle
27 Spine of a vertebra
28 Semispinalis capitis muscle

1 Submental lymph nodes
2 Anterior jugular lymph nodes
3 Submandibular lymph nodes
4 Paramandibular lymph nodes
5 Lateral retropharyngeal lymph nodes
6 Superficial lateral cervical lymph nodes
7 Inferior jugular lymph nodes
8 Medial retropharyngeal lymph nodes
9 Deep cervical lymph nodes
10 Nuchal lymph nodes

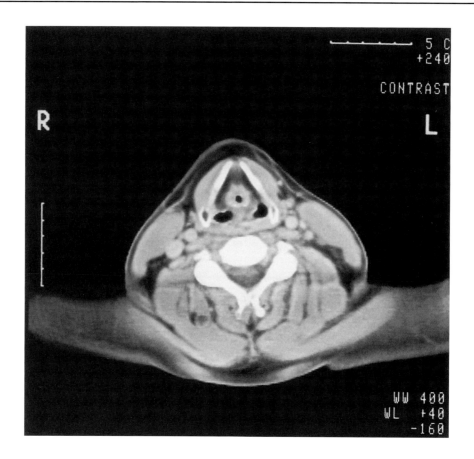

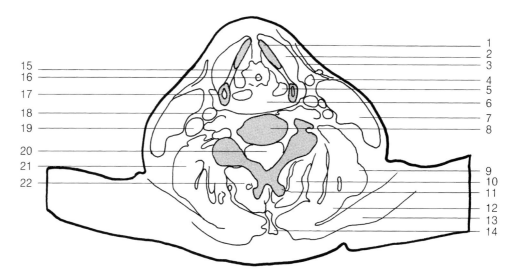

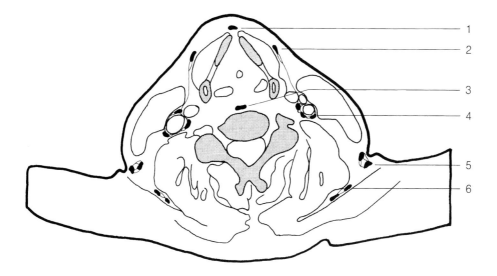

Fig. 16. Axial section of the neck   ▲

◄   1 Thyroid cartilage
    2 Sternohyoid, omohyoid and
      thyrohyoid muscles
    3 Platysma
    4 Thyroid artery and vein
    5 Piriform fossa
    6 Inferior constrictor muscle
      of the pharynx
    7 Sternocleidomastoid muscle
    8 Cervical vertebra
    9 Levator muscle of scapula
   10 Semispinalis cervicis muscle
   11 Spine of a vertebra
   12 Semispinalis capitis muscle
   13 Trapezius muscle
   14 Nuchal ligament
   15 Elastic membrane of larynx
   16 Cavity of larynx
   17 Superior horn of thyroid cartilage
   18 Internal carotid artery
   19 Internal jugular vein
   20 Vertebral canal
   21 Scalene muscles, costocervical muscle,
      levator muscle of scapula
   22 Multifidus muscle

   1 Prelaryngeal lymph nodes
   2 Anterior jugular lymph nodes
   3 Medial retropharyngeal lymph nodes
   4 Deep inferior jugular lymph nodes
   5 Superficial cervical lymph nodes
   6 Nuchal lymph nodes

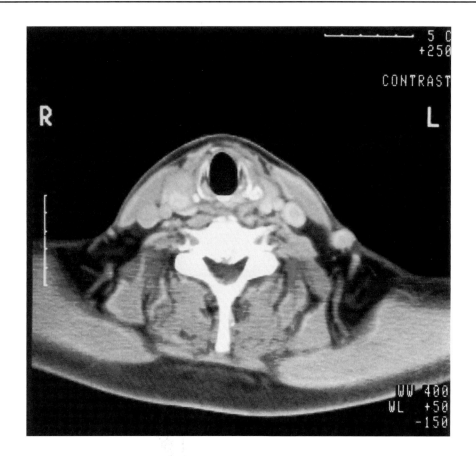

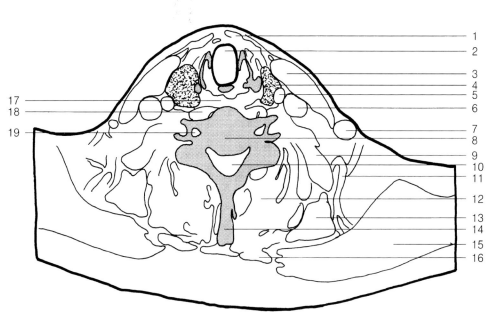

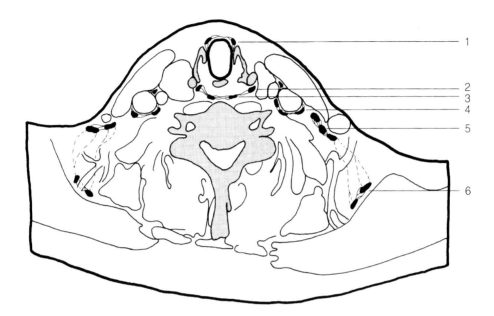

Fig. 17. Axial section of the neck ▲

◄ 1 Sternohyoid muscle
  2 Trachea
  3 Sternocleidomastoid muscle
  4 Thyroid gland
  5 Common carotid artery
  6 Internal jugular vein
  7 External jugular vein
  8 Cervical vertebra
  9 Scalene muscles
10 Vertebral canal
11 Transverse cervical vein
12 Splenius capitis and cervicis muscles
13 Supraspinous muscle
14 Spine of a vertebra
15 Trapezius muscle
16 Rhomboid muscle
17 Esophagus
18 Long muscle of neck
19 Transverse foramen

1 Pretracheal lymph nodes
2 Paratracheal lymph nodes
3 Medial retropharyngeal lymph nodes
4 Deep inferior jugular lymph nodes
5 Supraclavicular lymph nodes
6 Superficial cervical lymph nodes

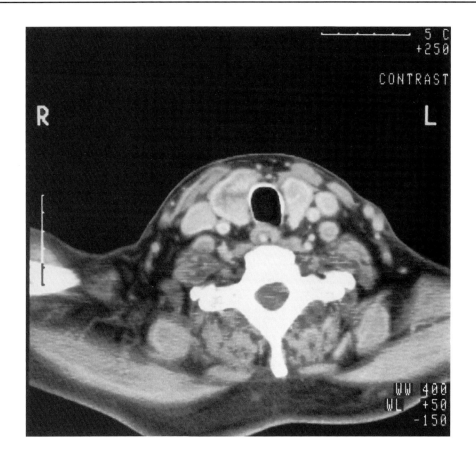

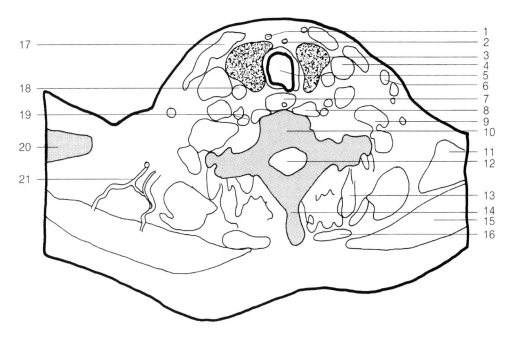

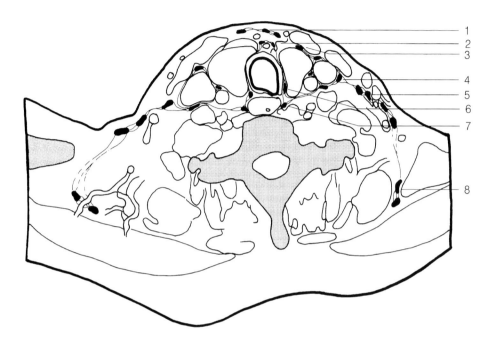

Fig. 18. Axial section of the neck ▲

◄ 1 Anterior jugular vein
   2 Thyroid vein
   3 Thyroid gland
   4 Internal jugular vein
   5 Trachea
   6 External jugular vein
   7 Esophagus
   8 Long muscle of neck
   9 Anterior and middle scalene muscles
  10 Cervical vertebra
  11 Supraspinous muscle
  12 Vertebral canal
  13 Splenius cervicis and capitis muscles
  14 Spine of a vertebra
  15 Trapezius muscle
  16 Rhomboid muscle
  17 Sternocleidomastoid muscle
  18 Common carotid artery
  19 Vertebral and superior thyroid artery
     and vein
  20 Clavicle
  21 Transverse cervical vein

   1 Anterior cervical lymph nodes
   2 Pretracheal lymph nodes
   3 Thyroid lymph node
   4 Inferior jugular lymph nodes
   5 Supraclavicular lymph nodes
   6 Paratracheal lymph nodes
   7 Retropharyngeal lymph nodes
   8 Superficial cervical lymph nodes

Thorax and Breast

General Considerations

In contrast to the lymphatic structures of other parts of the body, the intrathoracic lymphatic system is characterized by the black coloration resulting from the inhalation of dust and carbonaceous pigment. The embryonic origin of the peculiar intrathoracic pathways of lymphatic drainage was first studied by Sabin [43]. A lymphatic bud extends from the left jugular venous sac to form the lymphatics of the esophagus, the thoracic portion of the aorta, the terminal portion of the thoracic duct and the upper segments of the left lung. A lymphatic bud from the right jugular venous sac gives origin to the lymphatics of the heart and the remaining parts of the lung, except the basal segments of the lower lobes, whose lymphatics arise from the retroperitoneal lymph sac in the upper abdomen [10].

The efferent vessels of the mediastinal lymph node chains ascend into the cervical region and give rise to the bronchomediastinal trunk on each side. The left bronchomediastinal trunk terminates in the thoracic duct. On the right, the bronchomediastinal trunk inconstantly unites with the right subclavian trunk and the right jugular trunk to form the right lymphatic duct, which terminates in the venous angle between the subclavian and the internal jugular vein. The left subclavian trunk, which carries lymph from the upper extremity, accompanies the subclavian vein and terminates in the angle between the left subclavian vein and the internal jugular vein. The left jugular trunk carries lymph from the neck and terminates in the same place.

The thoracic duct arises from the cisterna chyli, ascends in the thoracic cavity anterior to the vertebral column and posterior to the aorta and drains into the angle of the left subclavian vein with the left jugular vein or, as a variant, into one of these two veins. Another variant is duplication of the upper part of the thoracic duct.

Lymphatic Drainage Regions

Lung (Fig. 19)

Each segment of the lung has its own separate lymphatic system. Two different lymphatic networks can be differentiated. The deep or central network originates in the connective tissue of the terminal bronchioles and accompanies the bronchi and the arterial blood vessels in the center of the lung segments. Especially in the region of the tra-

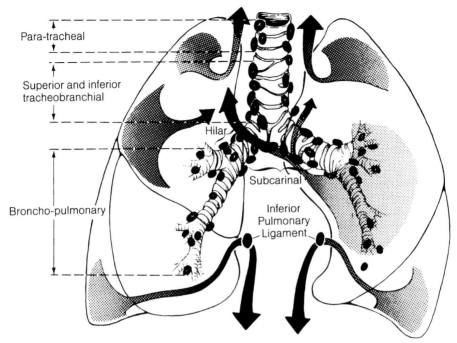

Para-tracheal

Superior and inferior
tracheobranchial

Hilar

Subcarinal

Inferior
Pulmonary
Ligament

Broncho-pulmonary

Fig. 19. Distribution of lymph nodes in relationship to the major bronchi and trachea. The *arrows* indicate the direction of the lymph flow to the mediastinum and extrathoracic sites. (Adapted from [13])

cheobronchial tree a rich lymph capillary network in the mucosa and submucosa can be demonstrated. The regional lymph nodes for the deep network are the bronchopulmonary nodes. These nodes can be found in the major angles of the bronchial tree adjoining the branches of the pulmonary arteries. The superficial or subpleural network originates in the subserous coat and accompanies the veins in the intersegmental and interlobular connective tissue septa. The regional lymph nodes for this network are the upper and lower tracheobronchial nodes.

Except for the lingula, the upper lobes are drained to the upper tracheobronchial nodes. The middle lobe, the lateroventral parts of the right lower lobe, the lingula and the lateral parts of the left lower lobe drain to those bronchopulmonary nodes which are situated between the middle and lower lobar bronchus on the right and between the upper and lower lobar bronchus on the left. However, lymph of the middle lobe and the lateroventral parts of the right lower lobe empties partly into the lower tracheobronchial nodes (Fig. 20). The lymph of the medial and dorsal parts of the lower lobes is collected by the lower tracheobronchial nodes. The lymph of the inferior surface of the lower lobes empties into nodes in the pulmonary ligament, thus communicating with subdiaphragmatic lymph vessels [43]. Additionally, the lower lobes drain into juxtaesophageal pulmonary nodes [31].

Two inconstant pathways of lymphatic drainage have to be mentioned: the upper parts of the right and left upper lobe may drain to the node at the azygos vein arch and to the node between the aortic arch and the left pulmonary artery (node of ligamentum arteriosum) respectively.

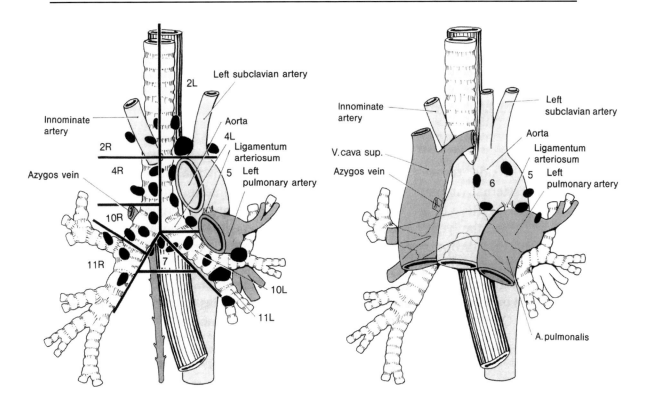

Fig. 20. Classification of the regional lymph nodes of the lung according to the American Thoracic Society [27]

2R + 2L Paratracheal lymph nodes
4R + 4L Superior tracheobronchial lymph nodes
5 (+6, not illustrated) anterior mediastinal
   lymph nodes
7 Subcorynal lymph nodes

8 + 9 (Not illustrated) posterior mediastinal
   lymph nodes
10R + 10L Bronchopulmonary lymph nodes
11 Pulmonary lymph nodes

The bronchopulmonary lymph nodes drain either directly or via the tracheo-bronchial nodes to the paratracheal nodes. The subsequent drainage is described in the chapter of the mediastinum.

The lymphatic drainage of both lungs is ipsilateral. However, lymph of the left lower lobe may cross over to the right via the lower tracheobronchial lymph nodes (Fig. 20; [15]).

Regional lymph nodes　　*Lnn. mediastinales posteriores*
(Lnn. tracheobronchiales superiores et inferiores
Lnn. paratracheales
Ln. arcus venae azygos
Lnn. bronchopulmonales)
Lnn. mediastinales anteriores
Ln. ligamentis arteriosi

Mediastinum

Following *Nomina Anatomica* [32] the mediastinal lymph nodes are divided into an anterior and a posterior group, though some authors refer to a third group, the intermediate or tracheobronchial group [9, 37, 42, 46].

Anterior Mediastinal Lymph Nodes

On the right side, these lymph nodes communicate with the right paratracheal nodes, parasternal (internal mammary) nodes, the left anterior mediastinal nodes and nodes lying close to the azygos vein at its entrance into the superior vena cava. On the left side, the lymph nodes of the anterior mediastinal chain have communications with the aortopulmonary, paratracheal, parasternal and anterior mediastinal nodes on the right side.

On the right side, the efferent vessels of the upper anterior mediastinal lymph nodes drain into the right lymphatic duct or into lymph nodes of the internal jugular vein group. On the left side, they empty into the thoracic duct.

Posterior Mediastinal Lymph Nodes (Fig. 21)

Paratracheal lymph nodes. The paratracheal group communicates with the left upper paraesophageal lymph nodes. The right paratracheal lymph nodes form the link between the superior tracheobronchial lymph nodes and the inferior deep cervical lymph nodes above, which may be found on the scalenus anterior muscle (scalene lymph nodes) [31].

Tracheobronchial lymph nodes. This group can be subdivided into the upper and lower tracheobronchial lymph nodes. The upper tracheobronchial lymph nodes include nodes at the termination of the azygos vein and nodes extending from the upper level of the bronchus to the summit of the aortic arch. The lower tracheobronchial lymph nodes consist of three clusters: the large subcorynal nodes at the bifurcation [46], the nodes of the lateral tracheobronchial angle and nodes lying lateral to the lowest part of the trachea.

The arrangement of the tracheobronchial lymph nodes on the left side differs from that on the right side mainly because of the presence of the aortic arch and the space between it and the left pulmonary artery, the so-called aortopulmonary window. The nodes within this space receive lymph from the upper segments of the left upper lobe, and their efferent lymph vessels communicate with the left upper paratracheal lymph nodes or empty into the thoracic duct or the left internal jugular or subclavian vein [42].

Juxtaesophageal pulmonary lymph nodes. These consist of lymph node chains which lie para-aortoesophageal and in juxtaposition to the pulmonary ligaments on both sides.

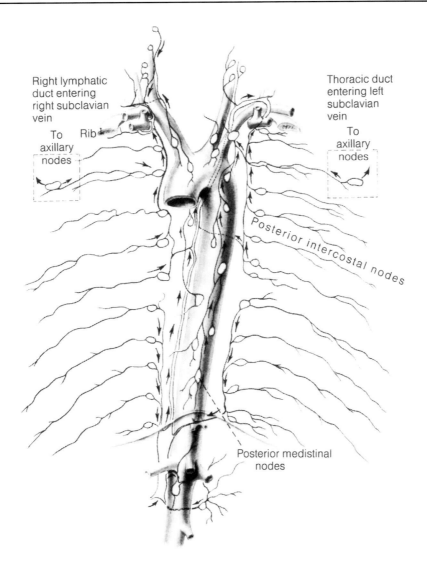

Fig. 21. Topography of the posterior mediastinal lymph nodes. The *arrows* mark the main directions of lymph flow. (Adapted from [15])

The lymph nodes of the ligaments can be divided into anterior nodes which drain the bases of the lower pulmonary lobes to the region of the cisterna chyli and posterior nodes which communicate with the thoracic duct. The upper nodes of the para-aorto-esophageal chain drain into paratracheal nodes or the thoracic duct, while the lower nodes empty into the cisterna chyli or the lower parts of the thoracic duct.

Regional lymph nodes   *Lnn. mediastinales anteriores*
           Lnn. mediastinales posteriores
           (Lnn. paratracheales
           Lnn. tracheobronchiales superiores et inferiores
           Lnn. juxta-esophageales pulmonales)

Heart and Pericardium

Developing from lymph vessels at the apex, one large lymph vessel in the longitudinal sulcus carries the lymph of the left ventricle to anterior mediastinal lymph nodes communicating with the right lymphatic duct or the thoracic duct. This vessel also receives lymph from the right ventricle via the right posterior main lymph vessels originating on the diaphragmatic surface of the heart. The lymph of the lateral and anterior parts of the pericardium is drained to the lateral pericardial and the prepericardial lymph nodes respectively.

Regional lymph nodes   *Lnn. mediastinales anteriores*
                          Lnn. pericardiales laterales
                          Lnn. prepericardiales

Thymus Gland

The upper thymic lymph vessels drain to an upper regional lymph node group medial to the junction of the subclavian and internal jugular veins. This group belongs to the tracheobronchial lymph nodes. The anterior thymic vessels empty into anterior mediastinal lymph nodes between the sternum and the thymus. The posterior thymic lymph nodes, part of the anterior mediastinal nodes, can be found between the pericardium and the thymus. The efferent vessels of the regional thymic lymph nodes open into the subclavian vein directly or unite with the subclavian, the jugular or the bronchomediastinal trunk.

Regional lymph nodes   *Lnn. mediastinales anteriores,*
                          Lnn. mediastinales posteriores
                          (Lnn. tracheobronchiales)

Esophagus (Fig. 22)

Resano's classification [391 divides the esophagus into four segments for description of lymph drainage: In the *supra-aortic segment* the anterolateral lymph vessels empty into nodes at the lower border of the inferior constrictor muscle of the pharynx whereas the posterolateral vessels terminate in nodes close to the inferior thyroid artery. Additionally, esophageal lymph may drain into retropharyngeal nodes and nodes in juxtaposition with the middle and lower thirds of the jugular vein.

The regional lymph nodes of the *retro-aortic segment* (zone of the aortic arch) can be found between the trachea and the esophagus on both sides (upper paratracheal group).

The regional lymph nodes of the *hilar segment* are situated between the esophagus and the descending aorta on the left side and in the pulmonary ligament on the right

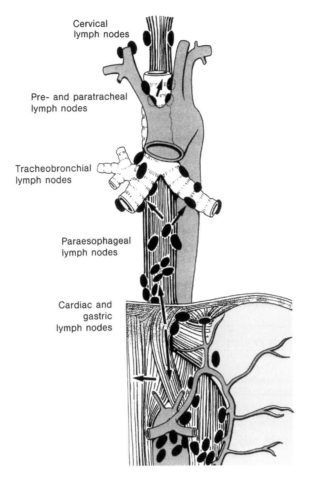

Fig. 22. Topography of the regional lymph nodes of the esophagus. The *arrows* indicate the main lymph flow directions. (Adapted from [17])

side (posterior mediastinal lymph nodes). The lymphatic flow of the hilar segment is directed both upward and downward, the latter to the celiac axis.

The *inferior segment*, i.e., that below the level of the pulmonary veins, drains to subdiaphragmatic lymph nodes (celiac axis lymph nodes).

Regional lymph nodes

Lnn. cervicales profundi
(Lnn. retropharyngeales,
Lnn. jugulares anteriores et laterales)
Lnn. mediastinales posteriores
(Lnn. paratracheales)
Lnn. coeliaci

Costal Wall and Parietal Pleura

The lymph is drained along the inferior border of the ribs to the parasternal (internal mammary) lymph nodes (anterolateral parts of intercostal spaces) or to the posterior intercostal lymph nodes (posterolateral intercostal spaces). At the first and second interspaces on the right, lymphatic communications may exist with a lymph node at the origin of the right subclavian artery, an anterior mediastinal lymph node at the termination of the azygos vein into the superior vena cava, and axillary lymph nodes (second and third interspaces). At the first two interspaces on the left, there are communications with a lymph node at the origin of the left common carotid artery, on the left innominate vein or on the internal jugular vein.

The posterior intercostal lymph nodes empty into collecting lymph vessels on each side of the vertebral bodies which terminate in the right lymphatic duct or in the thoracic duct or cisterna chyli.

The prevertebral lymph chains should also be mentioned as they are part of the parietal intrathoracic lymphatics.

Regional lymph nodes   *Lnn. parasternales*
Lnn. intercostales
Lnn. prevertebrales

Diaphragm

The lymph of the anterior part of the diaphragm (between pericardial sac and sternum) is drained to the lower parasternal lymph nodes. The lymph of the middle parts is collected by lymph nodes adjoining the phrenic nerves, while that of the posterior parts of the diaphragm empties into paraesophageal lymph nodes.

Regional lymph nodes   *Lnn. parasternales*
Lnn. phrenici superiores
Lnn. mediastinales posteriores

Breast (Fig. 23)

The lymph of the breast is drained mainly via the axillary and the internal mammary pathways. From a clinical point of view, the axillary lymph nodes (Lnn. axillares superficiales et profundi, Lnn. interpectorales) may be subdivided into six groups [42]:

1. The external mammary lymph nodes, also called pectoral lymph nodes, belonging to the group of the superficial axillary lymph nodes
2. The subscapular lymph nodes (Lnn. axillares superficiales)

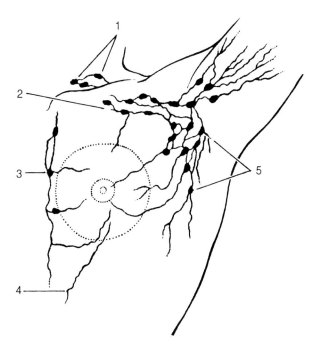

Fig. 23. Schematic illustration of the regional lymph nodes of the breast. (Adapted from [16])

1 Supraclavicular nodes
2 Apical axillary nodes
3 Parasternal node

4 Toward abdominal wall
5 Pectoral axillary nodes

3. The central lymph nodes (Lnn. axillares profundi), the largest and the most numerous axillary lymph nodes
4. The interpectoral lymph nodes, also called Rotter's nodes [40]
5. The axillary vein or lateral lymph nodes (Lnn. axillares superficiales)
6. The subclavicular or apical lymph nodes (Lnn. axillares profundi), receiving the collecting lymph vessels from all the other axillary lymph nodes

Additionally, a separate lymph node group may be identified at the lateral edge of the breast (Lnn. paramammarii).

It must be borne in mind that all axillary lymph nodes are connected by lymph vessels, the so-called axillary lymphatic plexus. The subclavian trunks may drain into either the venous angle or the so-called sentinel lymph nodes. The latter belong to the inferior deep cervical group and lie beneath the lateral and inferior part of the sternocleidomastoid muscle behind the clavicle. As a part of the transverse cervical chain the supraclavicular lymph nodes receive lymph from the axillary lymph nodes, either directly via the brachial plexus or indirectly via the sentinel lymph nodes.

The main lymphatic drainage pathway of the breast is to the axillary lymph nodes, but for the medial and deep parts of the gland the internal mammary chain is of crucial importance (Fig. 24). The lymph vessels in this region accompany blood vessels which perforate the pectoralis major muscle and reach the medial ends of the intercostal

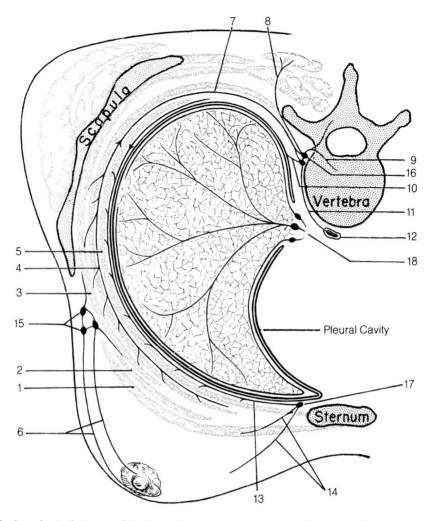

Fig. 24. The lymphatic drainage of the breast in cross section. (Adapted from [15])

 1 Greater pectoral muscle
 2 Smaller pectoral muscle
 3 Serratus anterior muscle
 4, 5 Intercostal muscle
 6 Lymphatics from breast to axillary
   lymph nodes
 7 External intercostal lymphatics
 8 Lymphatics from spinal muscles
 9 Lymphatics from vertebra
10 Lymphatics from parietal pleura

11 Lymphatics from posterior intercostal lymph
   nodes to thoracic duct
12 Thoracic duct
13 Internal intercostal lymphatics
14 Lymphatics from breast to internal
   mammary lymph nodes
15 Axillary lymph nodes
16 Posterior intercostal lymph nodes
17 Internal mammary lymph nodes
18 Lymph nodes of pulmonary pedicle

spaces, where they penetrate the intercostal muscles and terminate in the internal mammary lymph nodes (Lnn. parasternales). These lymph nodes are very small in diameter (2–5 mm) and can be found in close relationship to the internal mammary vessels within 3 cm of the sternal edge. At the first interspace level retromanubrial lymph nodes can be demonstrated in the majority of patients [2] thus connecting the left and right lymphatic trunk of the internal mammary chain.

Regional lymph nodes   *Lnn. axillares superficiales*
(Lnn. laterales, Lnn. pectorales, Lnn. subscapulares)
Lnn. axillares profundi
(Lnn. centrales, Lnn. apicales)
Lnn. axillares interpectorales
Lnn. paramammarii
Lnn. parasternales

Topographical Anatomy of Regional Lymph Node Groups of Mediastinum, Breast and Axillary Region (Figs. 25–37)

Lnn. Cervicales Profundi

Lnn. retropharyngeales (Figs. 5, 6, 11–18): behind the pharynx in front of the prevertebral fascia and muscles (Lnn. retropharyngeales mediales, Figs. 11–17) and along the internal carotid artery from the base of the skull down to the thoracic inlet (Lnn. retropharyngeales laterales, Figs. 14, 12, 14, 15)

Lnn. jugulares anteriores et laterales: anterior and lateral to the internal jugular vein.

Lnn. Axillares

Lnn. Axillares Superficiales (Figs. 27–30)

Lnn. laterales: along the lateral part of the axillary vein between the tendon of the latissimus dorsi muscle and the origin of the thoracoacromial vein.

Lnn. pectorales (Fig. 28): the superior group follows the course of the lateral thoracic artery at the level of the second and third intercostal spaces, while the inferior group presents less frequently at the level of the 4th–6th intercostal space beneath the lateral edge of the pectoralis major muscle and accompanying the lateral thoracic artery on the chest wall.

Lnn. subscapulares: in close proximity to the subscapular vessels, their thoracodorsal branches and the pectoralis minor muscle.

Lnn. Interpectorales (Figs. 26, 29, 30)

Between the pectoralis minor and major muscles along the pectoral branches of the thoracoacromial vessels.

Lnn. Axillares Profundi (Figs. 26, 27)

Lnn. centrales (Fig. 26): on the anterior aspect of the subscapular muscle and posterior to the origin of the pectoralis minor muscle.

Lnn. apicales: in the apex of the axilla where the axillary vein disappears beneath the tendon of the subclavius muscle.

Lnn. Paramammarii (Figs. 32–36)

These may be found at the lateral edge of the breast.

Lnn. Parasternales (Figs. 29–36)

These nodes lie adjacent to the internal mammary vessels in the interspaces between the costal cartilages covered by the endothoracic fascia.

Lnn. Intercostales (Figs. 26–37)

These may be found at the terminations of the external intercostal muscles.

Lnn. Mediastinales Anteriores (Figs. 25–34)

These lymph nodes can be found on the anterior surface of the great vessels of the upper half of the thorax and along the thoracic portion of the phrenic nerves. On the right side, the pretracheal lymph nodes are located between the superior vena cava, the trachea and the aorta. On the left side, lymph nodes may be found on the anterior aspect of the aortic arch at the origin of the common carotid artery or the subclavian artery.

Nodus Ligamentis Arteriosi (Fig. 31)

This is an inconstant node at the ligamentum arteriosum.

Lnn. Pericardiales Laterales (Figs. 36, 37)

These are located at the lateral aspects of the pericardium covered by the mediastinal pleura.

Lnn. Prepericardiales (Fig. 35–37)

These lie between the sternum and the pericardium.

Lnn. Mediastinales Posteriores

Lnn. Paratracheales (Fig. 25–30)

This group lies between the trachea and the superior vena cava.

Lnn. Tracheobronchiales

Lnn. tracheobronchiales superiores (Fig. 30): from a posterolateral position at the superior level of the aortic arch to an anterior position just above the bifurcation of the trachea.

Lnn. tracheobronchiales inferiores (Figs. 31–33): anteroinferior and posteroinferior to the tracheobronchial angle and the bifurcation and at the posterior wall of the lowest part of the trachea.

Lnn. Bronchopulmonales (Figs. 32–35)

These lymph nodes are situated in the major angles of the bronchial tree. On the right side they can be found ventral to the pulmonary artery, whereas on the left side they occupy a posterolateral position with the right pulmonary artery and its branches lying ventrally.

Lnn. Juxta-esophageales Pulmonales (Figs. 30–36)

These can be found para-aortoesophageal and in juxtaposition to the pulmonary ligaments.

Nodus Arcus Venae Azygos (Fig. 31)

This is an inconstant node at the point where the azygos vein forms an arch before terminating in the superior vena cava.

Lnn. Prevertebrales (Figs. 25–30, 33, 34, 36, 37)

These lymph nodes are situated on the anterior and lateral aspects of the vertebral bodies.

Lnn. Phrenici Superiores (Fig. 37)

These are located in juxtaposition to the phrenic nerves.

Lnn. Coeliaci

These are situated around the celiac axis.

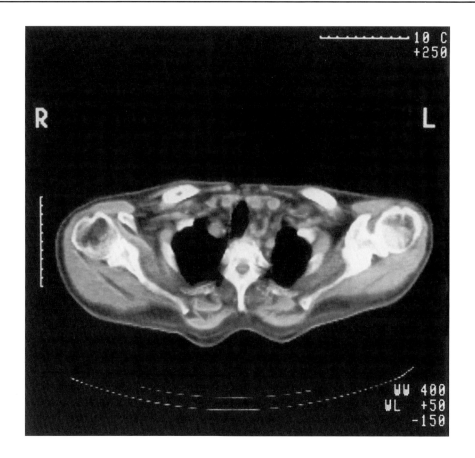

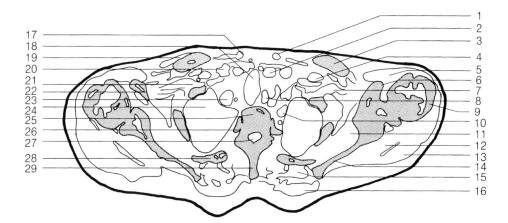

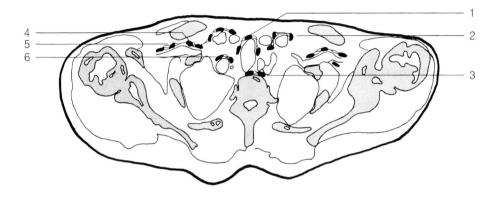

Fig. 25. Axial section of the thorax
(male subject, arms not elevated)   ▲

◄ 1 Anterior jugular vein
   2 Common carotid artery
   3 Clavicle
   4 Subclavian artery and vein
   5 Deltoid muscle
   6 Coracoid process
   7 Axillary vein
   8 Esophagus
   9 Head of humerus
  10 Subclavian artery
  11 Subscapular muscle
  12 Infraspinous muscle
  13 Deltoid muscle
  14 Scapula
  15 Rhomboid muscles
  16 Trapezius muscle
  17 Sternocleidomastoid muscle
  18 Trachea
  19 Greater pectoral muscle
  20 Internal jugular vein
  21 Subclavian artery and vein
  22 Coracoid process
  23 Innominate vein
  24 Lung
  25 Vertebra
  26 Head of a rib
  27 Vertebral canal
  28 Shaft of a rib
  29 Spine of a vertebra

1 Paratracheal lymph nodes
2 Anterior and lateral cervical lymph nodes
3 Prevertebral lymph nodes
4 Apical lymph nodes
5 Central lymph nodes
6 Anterior mediastinal lymph nodes

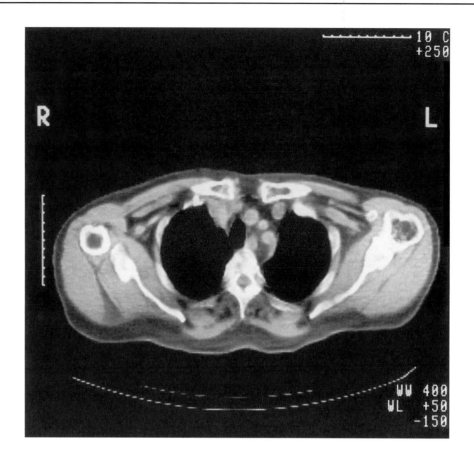

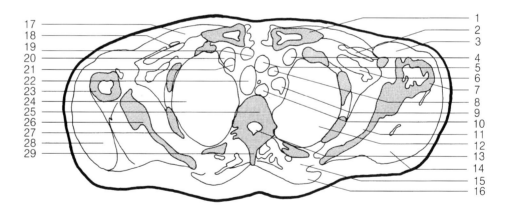

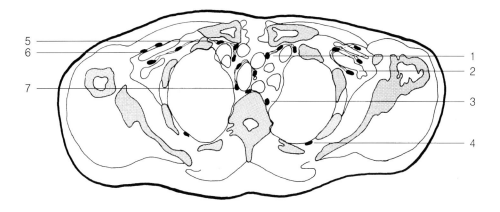

Fig. 26. Axial section of the thorax (male subject, arms not elevated)

◄

▲

 1 Clavicle
 2 Coracobrachial muscle
 3 Deltoid muscle
 4 Subclavian vein
 5 Innominate vein
 6 Axillary artery and vein
 7 Head of humerus
 8 Common carotid artery
 9 Subclavian artery
10 Esophagus
11 Subscapular muscle
12 Lung
13 Vertebral canal
14 Infraspinous muscle
15 Rhomboid muscles
16 Trapezius muscle
17 Greater pectoral muscle
18 Smaller pectoral muscle
19 Innominate artery
20 Right venous angle
21 Innominate artery
22 Trachea
23 Humerus
24 Lung
25 Vertebra
26 Scapula
27 Rib
28 Deltoid muscle
29 Spine of a vertebra

1 Anterior mediastinal lymph nodes
2 Central lymph nodes
3 Prevertebral lymph nodes
4 Intercostal lymph nodes
5 Anterior mediastinal lymph nodes
6 Interpectoral lymph nodes
7 Paratracheal lymph nodes

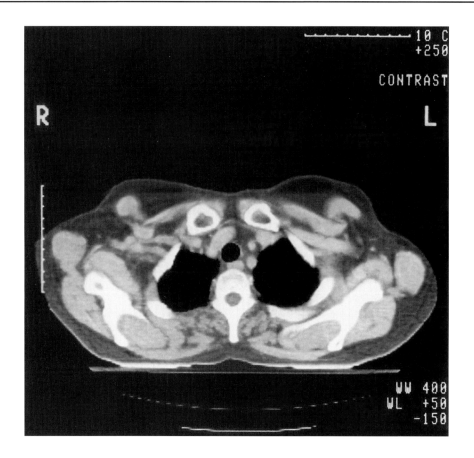

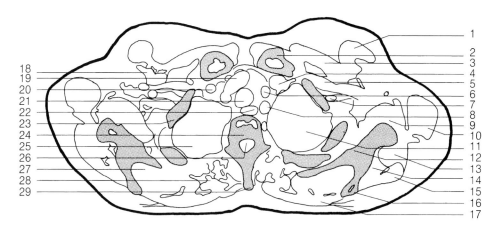

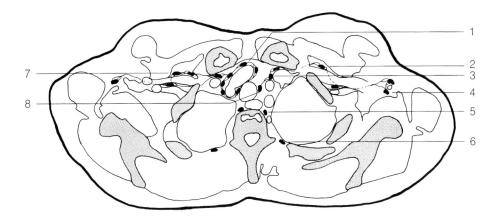

Fig. 27. Axial section of the thorax (female subject)

▲

◀ 1 Greater pectoral muscle
2 Clavicle
3 Greater pectoral muscle
4 Innominate vein
5 Axillary vein
6 Common carotid artery
7 Axillary artery
8 Subclavian artery
9 Subscapular muscle
10 Teres major muscle
11 Lung
12 Scapula
13 Teres minor muscle
14 Infraspinous muscle
15 Deltoid muscle
16 Spine of scapula
17 Trapezius muscle
18 Smaller pectoralis muscle
19 Innominate artery
20 Innominate vein
21 Trachea
22 Esophagus
23 Rib
24 Vertebra
25 Lung
26 Vertebral canal
27 Supraspinous muscle
28 Rhomboid muscles
29 Spine of a vertebra

1 Anterior mediastinal lymph nodes
2 Deep axillary lymph nodes
3 Anterior mediastinal lymph nodes
4 Superficial axillary lymph nodes
5 Prevertebral lymph nodes
6 Intercostal lymph nodes
7 Deep axillary lymph nodes
8 Paratracheal lymph nodes

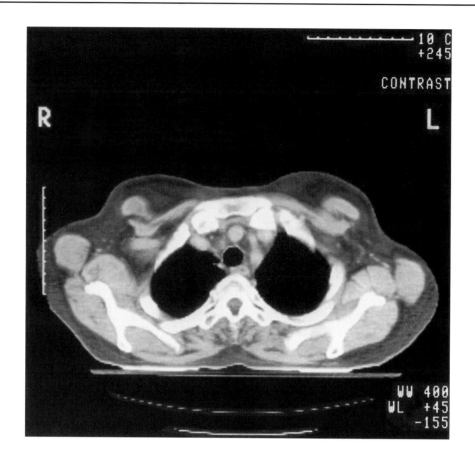

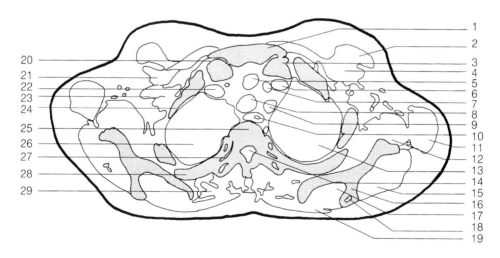

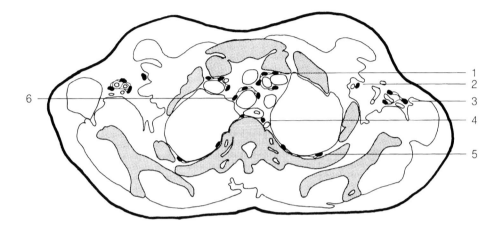

Fig. 28. Axial section of the thorax
(male subject)

▲

◀ 1 Manubrium sterni
  2 Greater pectoral muscle
  3 Clavicle
  4 Smaller pectoral muscle
  5 Innominate artery
  6 Innominate vein
  7 Left common carotid artery
  8 Trachea
  9 Subclavian artery
 10 Esophagus
 11 Teres major muscle
 12 Subscapular muscle
 13 Lung
 14 Vertebral canal
 15 Infraspinous muscle
 16 Scapula
 17 Supraspinous muscle
 18 Spine of scapula
 19 Trapezius muscle
 20 Sternoclavicular joint
 21 Smaller pectoral muscle
 22 Innominate vein
 23 Axillary vein
 24 Rib
 25 Vertebra
 26 Lung
 27 Costovertebral joint
 28 Rib
 29 Rhomboid muscles

1 Anterior mediastinal lymph nodes
2 Deep axillary lymph nodes (pectoral lymph nodes)
3 Superficial axillary lymph nodes
4 Prevertebral lymph nodes
5 Intercostal lymph nodes
6 Paratracheal lymph nodes

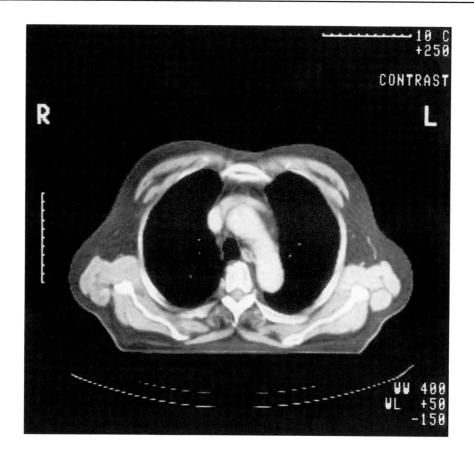

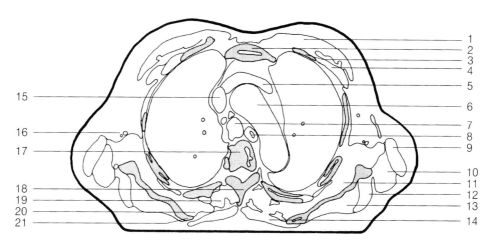

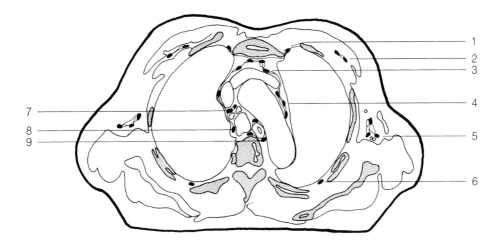

Fig. 29. Axial section of the thorax (male subject)

▲

 1 Greater pectoral muscle
 2 Sternum
 3 Rib
 4 Smaller pectoral muscle
 5 Innominate vein
 6 Arch of aorta
 7 Trachea
 8 Esophagus
 9 Latissimus dorsi muscle
10 Teres major muscle
11 Subscapular muscle
12 Teres minor muscle
13 Infraspinous muscle
14 Trapezius muscle
15 Superior vena cava
16 Lateral thoracic artery and vein
17 Vertebra
18 Spine of a vertebra
19 Erector muscle of spine
20 Rhomboid muscles
21 Scapula

 1 Parasternal lymph nodes
 2 Interpectoral axillary lymph nodes
 3, 4 Anterior mediastinal lymph nodes
 5 Superficial axillary lymph nodes
 6 Intercostal lymph nodes
 7 Anterior mediastinal lymph nodes
 8 Paratracheal lymph nodes
 9 Prevertebral lymph nodes

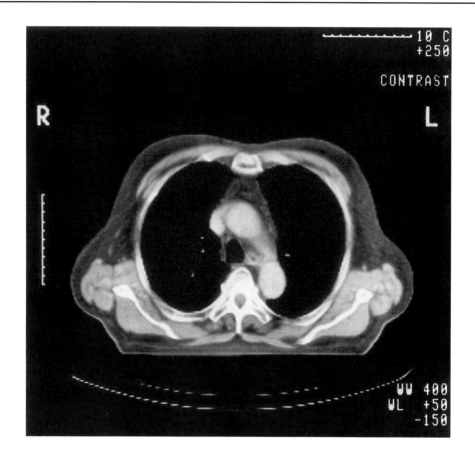

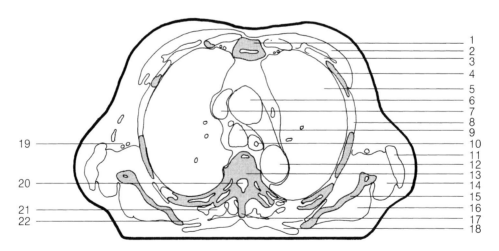

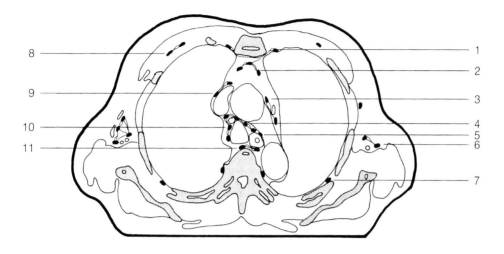

Fig. 30. Axial section of the thorax
(male subject)   ▲

◄ 1 Sternum
  2 Greater pectoral muscle
  3 Smaller pectoral muscle
  4 Rib
  5 Lung
  6 Ascending aorta
  7 Superior vena cava
  8 Intercostal muscles
  9 Trachea
10 Esophagus
11 Latissimus dorsi muscle
12 Descending aorta
13 Vertebra
14 Teres major muscle
15 Scapula
16 Infraspinous muscle
17 Spine of a vertebra
18 Trapezius muscle
19 Lateral thoracic artery and vein
20 Vertebral canal
21 Erector muscle of spine
22 Rhomboid muscles

  1 Parasternal lymph nodes
2, 3 Anterior mediastinal lymph nodes
  4 Superior tracheobronchial lymph nodes
  5 Juxtaesophageal pulmonary lymph nodes
  6 Superficial axillary lymph nodes
  7 Intercostal lymph nodes
  8 Interpectoral axillary lymph nodes
  9 Anterior mediastinal lymph nodes
10 Paratracheal lymph nodes
11 Prevertebral lymph nodes

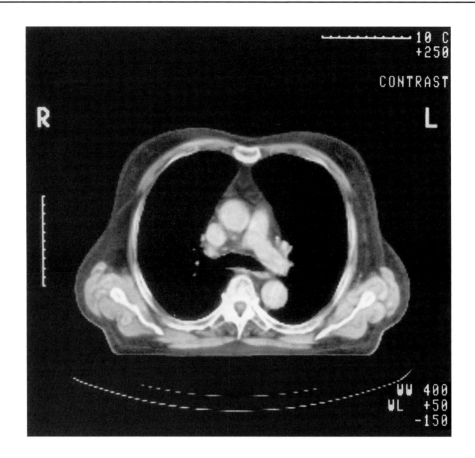

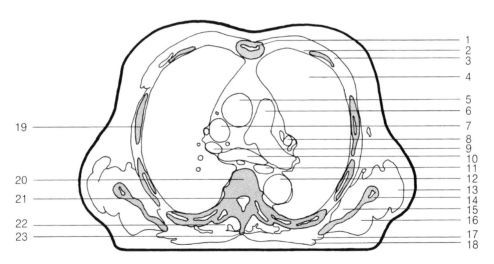

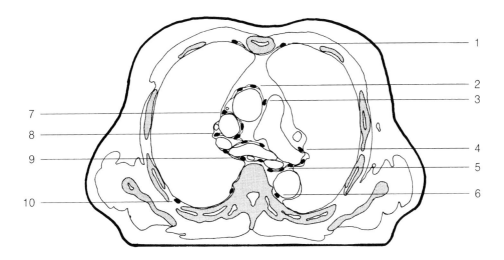

Fig. 31. Axial section of the thorax
(male subject)

▲

◀ 1 Sternum
  2 Costal cartilage
  3 Greater pectoral muscle
  4 Lung
  5 Ascending aorta
  6 Pulmonary trunk
  7 Superior vena cava
  8 Left superior pulmonary vein
  9 Right pulmonary artery
10 Main bronchus
11 Esophagus
12 Descending aorta
13 Latissimus dorsi muscle
14 Scapula
15 Subscapular muscle
16 Infraspinous muscle
17 Rhomboid muscles
18 Trapezius muscle
19 Rib
20 Vertebra
21 Vertebral canal
22 Erector muscle of spine
23 Spine of a vertebra

  1 Parasternal lymph nodes
  2 Anterior mediastinal lymph nodes
  3 Node of ligamentum arteriosum
  5 Juxtaesophageal pulmonary lymph nodes
  6 Intercostal lymph nodes
  7 Anterior mediastinal lymph nodes
  8 Node of arch of azygos vein
  9 Inferior tracheobranchial lymph nodes
10 Intercostal lymph nodes

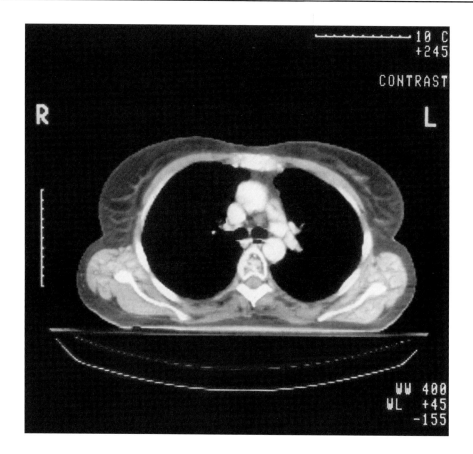

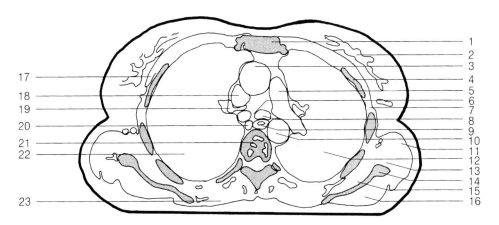

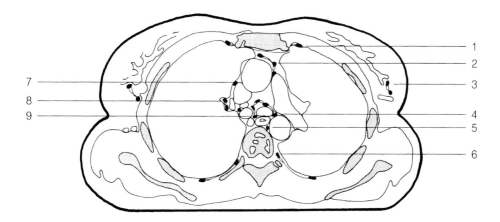

Fig. 32. Axial section of the thorax
(female subject)

▶ ◀

 1 Sternum
 2 Body of breast
 3 Ascending aorta
 4 Rib
 5 Pulmonary trunk
 6 Carina
 7 Intercostal muscles
 8 Main bronchus
 9 Esophagus
10 Descending aorta
11 Latissimus dorsi muscle
12 Lung
13 Teres major muscle
14 Scapula
15 Subscapular muscle
16 Infraspinous muscle
17 Greater pectoral muscle
18 Superior vena cava
19 Apical branch of right pulmonary artery
20 Azygos vein
21 Vertebra
22 Lung
23 Trapezius muscle

▲

1 Parasternal lymph nodes
2 Anterior mediastinal lymph nodes
3 Paramammary lymph nodes
4 Inferior tracheobronchial lymph nodes
5 Juxtaesophageal pulmonary lymph nodes
6 Intercostal lymph nodes
7 Anterior mediastinal lymph nodes
8 Bronchopulmonary lymph nodes
9 Inferior tracheobronchial lymph nodes

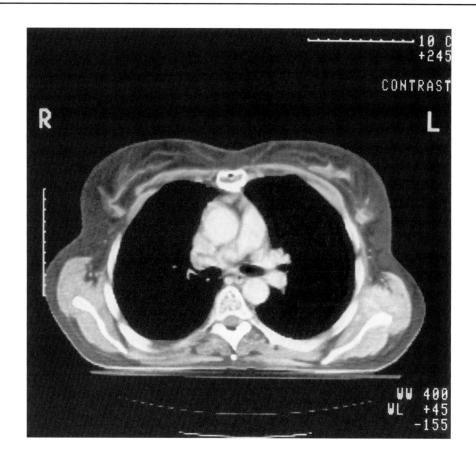

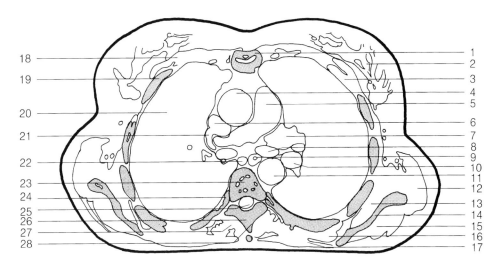

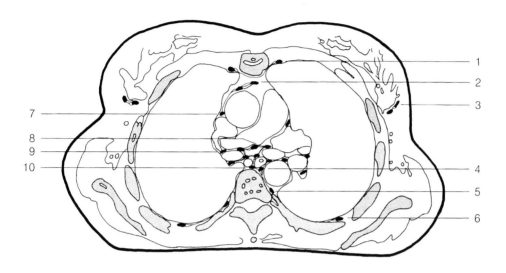

Fig. 33. Axial section of the thorax
(female subject)

▲

◄ 1 Sternum
    2 Body of breast
    3 Costal cartilage
    4 Ascending aorta
    5 Pulmonary trunk
    6 Superior vena cava
    7 Pulmonary vein
    8 Main bronchus
    9 Esophagus
   10 Pulmonary artery
   11 Descending aorta
   12 Latissimus dorsi muscle
   13 Subscapular muscle
   14 Scapula
   15 Infraspinous muscle
   16 Rhomboid muscles
   17 Trapezius muscle
   18 Greater and smaller pectoral muscles
   19 Rib
   20 Lung
   21 Right pulmonary vein
   22 Azygos vein
   23 Vertebra
   24 Head of a rib
   25 Vertebral canal
   26 Vertebral arch
   27 Erector muscle of spine
   28 Spine of a vertebra

    1 Parasternal lymph nodes
    2 Anterior mediastinal lymph nodes
    3 Paramammary lymph nodes
    4 Juxtaesophageal pulmonary lymph nodes
 5, 6 Intercostal lymph nodes
    7 Anterior mediastinal lymph nodes
    8 Inferior tracheobronchial lymph nodes
    9 Bronchopulmonary lymph nodes
   10 Prevertebral lymph nodes

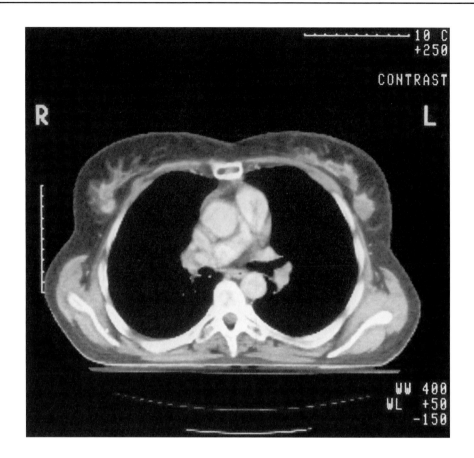

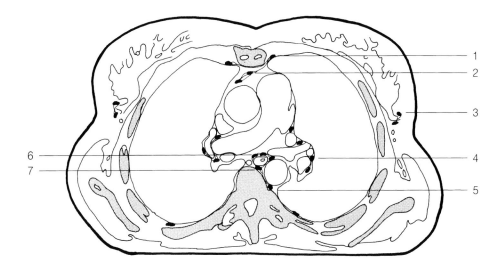

Fig. 34. Axial section of the thorax (female subject)

▲

◄ 1 Sternum
  2 Body of breast
  3 Pulmonary trunk
  4 Ascending aorta
  5 Rib
  6 Lung
  7 Pulmonary vein
  8 Lobar bronchus
  9 Main bronchus
10 Esophagus
11 Left pulmonary artery
12 Descending aorta
13 Scapula
14 Infraspinous muscle
15 Trapezius muscle
16 Superior versa cava
17 Right pulmonary vein
18 Latissimus dorsi muscle
19 Main bronchus
20 Azygos vein
21 Serratus anterior muscle
22 Vertebra
23 Vertebral canal
24 Erector muscle of spine
25 Spine of a vertebra

1 Parasternal lymph nodes
2 Anterior mediastinal lymph nodes
3 Paramammary lymph nodes
4 Bronchopulmonary lymph nodes
5 Intercostal lymph nodes
6 Juxtaesophageal pulmonary lymph nodes
7 Prevertebral lymph nodes

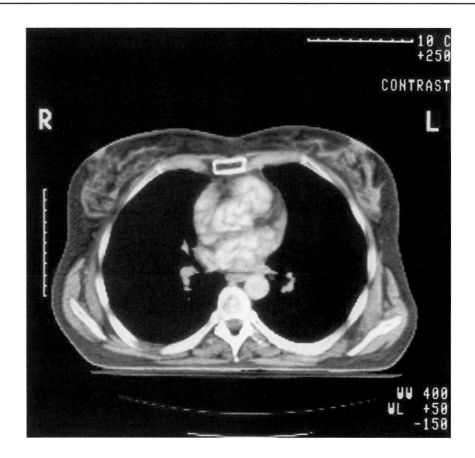

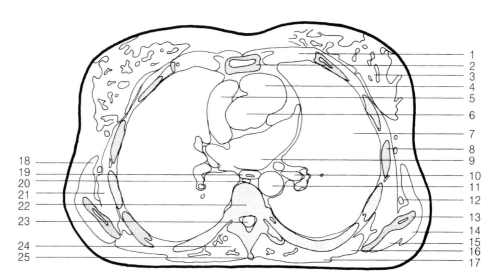

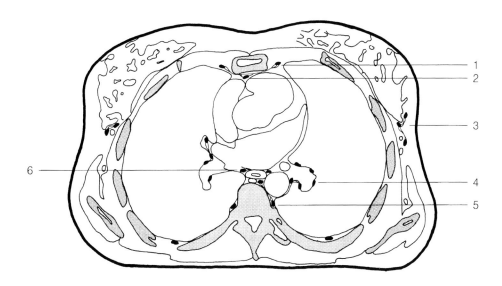

Fig. 35. Axial section of the thorax
(female subject)

▲

◄ 1 Costal cartilage
   2 Sternum
   3 Body of breast
   4 Right ventricle
   5 Right atrium
   6 Ascending aorta
   7 Lung
   8 Rib
   9 Left atrium
  10 Lobar bronchus
  11 Descending aorta
  12 Thoracodorsal artery and vein
  13 Scapula
  14 Teres major muscle
  15 Infraspinous muscle
  16 Rhomboid muscles
  17 Trapezius muscle
  18 Latissimus dorsi muscle
  19 Esophagus
  20 Azygos vein
  21 Serratus anterior muscle
  22 Vertebra
  23 Vertebral canal
  24 Erector muscle of spine
  25 Spine of a vertebra

  1 Parasternal lymph nodes
  2 Prepericardial lymph nodes
  3 Paramammary lymph nodes
  4 Bronchopulmonary (hilar) lymph nodes
  5 Intercostal lymph nodes
  6 Juxtaesophageal pulmonary lymph nodes

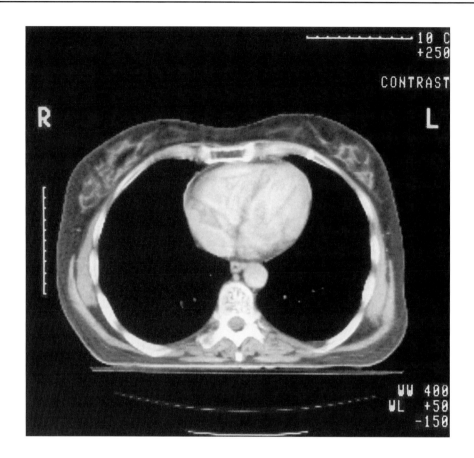

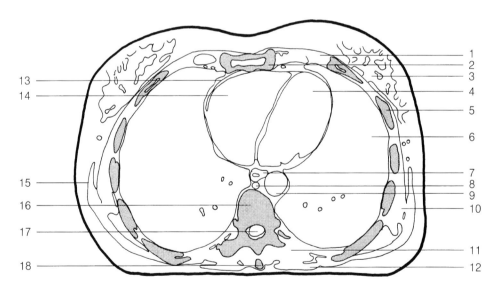

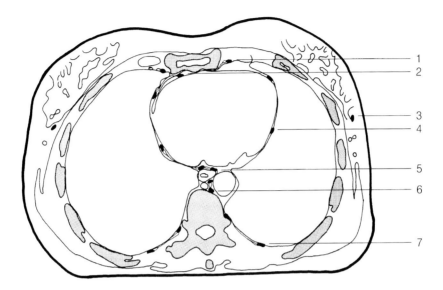

Fig. 36. Axial section of the thorax
(female subject)

▲

◀ 1 Costal cartilage
2 Sternum
3 Body of breast
4 Left ventricle
5 Rib
6 Lung
7 Esophagus
8 Azygos vein
9 Descending aorta
10 Serratus anterior muscle
11 Erector muscle of spine
12 Trapezius muscle
13 Pericardium
14 Right ventricle
15 Latissimus dorsi muscle
17 Vertebral canal
18 Spine of a vertebra

1 Parasternal lymph nodes
2 Prepericardial lymph nodes
3 Paramammary lymph nodes
4 Lateral pericardial lymph nodes
5 Juxtaesophageal pulmonary lymph nodes
6 Prevertebral lymph nodes
7 Intercostal lymph nodes

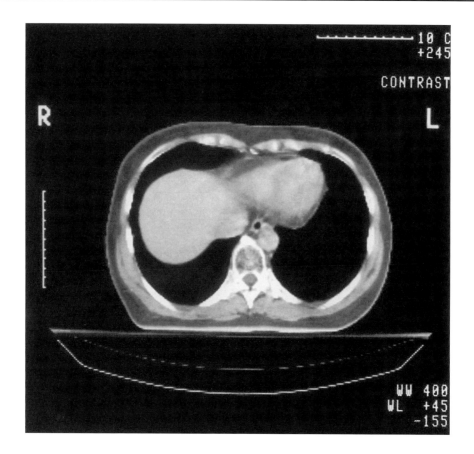

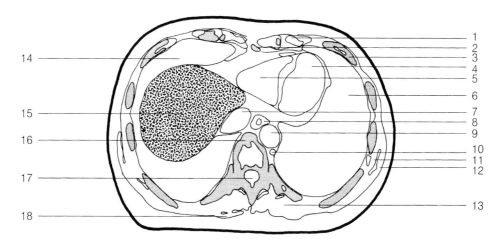

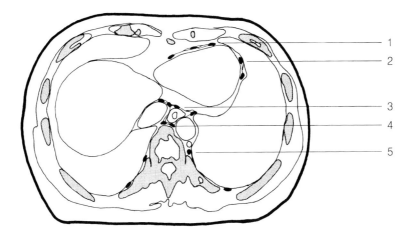

Fig. 37. Axial section of the thorax
(female subject)

▲

◄ 1 Costal cartilage
  2 Xiphoid process
  3 Rib
  4 Left ventricle
  5 Right ventricle
  6 Lung
  7 Inferior vena cava
  8 Esophagus
  9 Descending aorta
 10 Inferior hemiazygos vein
 11 Serratus anterior muscle
 12 Latissimus dorsi muscle
 13 Erector muscle of spine
 14 Lung
 15 Liver
 16 Vertebra
 17 Vertebral canal
 18 Trapezius muscle

1 Prepericardial lymph nodes
2 Lateral pericardial lymph nodes
3 Superior phrenic lymph nodes
4 Prevertebral lymph nodes
5 Intercostal lymph nodes

Abdomen

General Considerations (Fig. 38)

Retroperitoneal organs, e.g., kidneys and adrenal glands, and organs developing intraperitoneally, e.g., liver, spleen and pancreas, possess different lymphatic drainage areas. Whereas retroperitoneal organs drain mainly to the lumbar lymph node group, intraperitoneal organs drain to visceral nodes, e.g., the mesenteric lymph nodes, the largest group in the body comprising some 100–150 nodes. The mesenteric nodes drain the

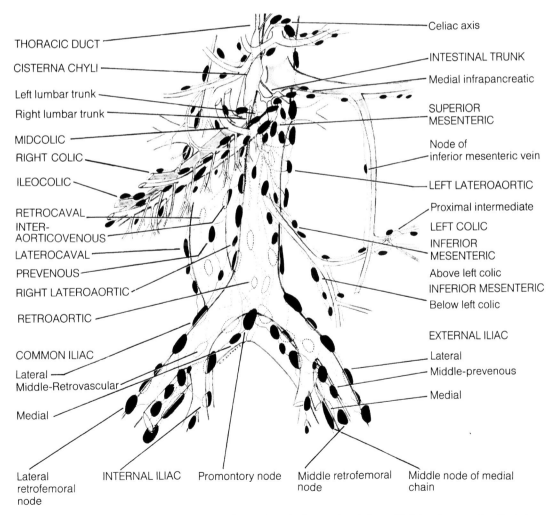

Fig. 38. Main lymph node regions of the abdomen in relation to the right and left lumbar trunks, the intestinal trunks and the cisterna chyli. (Adapted from [15])

small and large intestines except for the lower part of the descending colon and the sigmoid (via inferior mesenteric nodes to lateral aortic nodes) and the pelvic (para-aortic and internal iliac nodes) and anal (inguinal nodes) parts of the rectum. The efferent vessels of the mesenteric nodes form the intestinal trunk which, together with the lumbar trunks as the efferent lymph vessels of the lumbar chain, gives rise to the thoracic duct at the cisterna chyli.

The gastric and hepatic lymph nodes are located intraperitoneally. Due to organ rotation in the course of embryological development, the pancreaticosplenic lymph nodes secondarily take a retroperitoneal position but still communicate with central mesenteric and para-aortic nodes [20].

In fact, there are communications between the lymphatics of the abdomen and the thorax, e. g., efferent lymph vessels of the liver perforate the diaphragm and terminate in the anterior and posterior mediastinal lymph nodes.

Lymphatic Drainage Regions

Abdominal Wall

Two groups of regional lymph nodes can be demonstrated. The inconstant superficial group drains either to the paramammary or the subscapular nodes of the superficial axillary chain or to the medial and lateral superficial inguinal nodes. The umbilicus represents the watershed between the two territories.

The second group of regional nodes drains the deep lymphatics, especially those of the muscles of the abdominal wall, via several pathways: first the inferior epigastric chain, consisting of three to six nodes, second the internal mammarian chain and third via nodes along the deep circumflex iliac artery. Two further pathways consist of lymph vessels along the intercostal blood vessels and along the lumbar arteries, the latter terminating in lumbar nodes.

Regional lymph nodes   *Lnn. epigastrici inferiores*
                        Lnn. lumbales
                        Lnn. intercostales
                        Lnn. parasternales
                        Lnn. paramammarii
                        Lnn. axillares superficiales
                        (Lnn. subscapulares)

Liver (Fig. 39)

The superficial lymphatics of the liver arise from the interlobular spaces. Over the superior surface of the right lobe, they may be divided into anterior, superior and posterior collecting lymph vessels. The anterior vessels drain to the lymph nodes in the hepatic

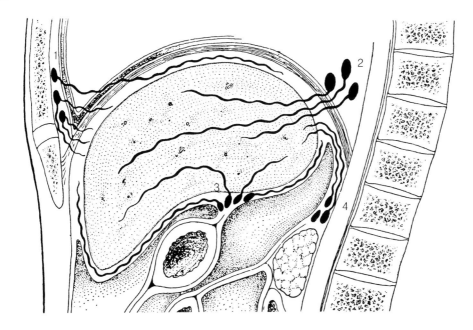

Fig. 39. Superficial and deep lymphatics of the liver draining to parasternal and pericardial nodes (1), posterior mediastinal nodes (2), hepatic nodes (3) and nodes at the celiac axis (4). (Adapted from [1])

pedicle. The superior vessels, lying in the coronary ligament, drain either to lymph nodes in the hepatic pedicle and around the inferior vena cava or, passing through the diaphragm, to parasternal and pericardial lymph nodes. The lymph of the posterior lymph vessels is carried to lymph nodes around the celiac artery trunk and the superior mesenteric artery or to posterior mediastinal lymph nodes.

As for the superior surface of the left lobe, the superior lymph vessels terminate in the superior collecting vessels of the right lobe, while the posterior lymph vessels penetrate the triangular ligament and run to the coronary chain and coeliac lymph nodes. At the inferior surface of the liver, the most posterior lymph vessels of the right lobe drain to para-aortic nodes, whereas all the others drain to nodes in the hepatic hilus and the hepatic pedicle. The latter also applies for the lymph vessels of the left lobe.

As for the deep lymphatics, some follow the branches of the portal vein and the hepatic artery and terminate in lymph nodes of the hilus, the hepatic pedicle, the hepatic artery and the coronary chain. Others drain to supradiaphragmatic lymph nodes around the inferior vena cava.

Regional lymph nodes	*Lnn. hepatici*
	Lnn. coeliaci
	Lnn. lumbales
	Lnn. parasternales
	Lnn. phrenici superiores
	Lnn. pericardiales laterales
	Lnn. mediastinales posteriores

Gallbladder and Extrahepatic Bile Ducts

The lymphatics of the gallbladder arise from the mucosa, penetrate the muscular wall and empty into a subserous lymphatic network. They anastomose richly with the hepatic lymphatics.

Four groups of collecting lymph vessels may be demonstrated [5]: two superior, intrahepatic groups (medial and lateral) and two inferior, extrahepatic groups (medial and lateral). The superior groups are drained via the cystic node at the neck of the gallbladder and the nodes in the hepatoduodenal ligament and around the common hepatic artery to para-aortic lymph nodes lying superior to the left renal vein. The lymph of the inferior groups is carried to the cystic node, the nodes in the hepatoduodenal ligament, the posterior pancreaticoduodenal nodes and the pre- and paraaortic nodes lying inferior to the left renal vein.

The lymphatics of the hepatic, cystic and common bile ducts comprise a rich network in the mucosa and on the surface. The lymph of the cystic duct is drained to the cystic node and the node of the epiploic foramen. The latter also receives lymph from the hepatic duct and the common bile duct, whose lymph vessels drain to pancreaticoduodenal nodes as well. The region of the hepatopancreatic ampulla shows lymphatic connections to the network of the duodenum and the pancreas.

Regional lymph nodes   *Lnn. hepatici*
                              (Nodus cysticus, Nodus foraminalis)
                              Lnn. lumbales
                              (Lnn. aortici laterales, Lnn. pre-aortici)
                              Lnn. lumbales intermedii

Stomach (Fig. 40)

The subserosal lymphatics collect the lymph of the mucosal, submucosal and intermuscular lymphatics and are drained by lymphatic vessels which accompany the blood vessels [16]. The lesser curvature is drained by vessels accompanying the left gastric artery in the lesser omentum. The lymph of the cardia is partly drained to lymph nodes of the celiac axis, paraesophageal nodes and lumbar lymph nodes which are grouped close to the left renal vein.

However, lymph from the most inferior part of the lesser curvature, the angular notch, the pylorus and the greater curvature is drained to nodes along the hepatic artery, the gastroduodenal and the gastroepiploic vessels. The remaining area, i. e., from the fundus to the first 2 cm of the greater curvature, has lymphatic vessels located in the gastrocolic and gastrosplenic ligaments along the course of the left gastroepiploic vessels. These lymph vessels drain to collecting vessels accompanying the splenic artery and vein.

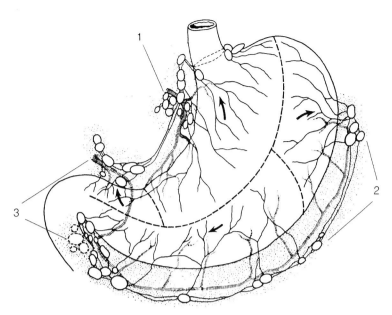

Fig. 40. Main lymphatic drainage regions of the stomach and their regional lymph nodes. (Adapted from [6])

1 Right and left gastric lymph nodes
2 Gastroomental lymph nodes
3 Pyloric lymph nodes

In living man, each side of the stomach may be divided into three longitudinal areas. As the intermediate area does not dispose of lymph vessels with valves, the lymph flows to the area of either the lesser or the greater curvature [35].

Regional lymph nodes   *Lnn. coeliaci*
                          Lnn. gastrici dextri et sinistri
                          Lnn. gastro-omentales dextri et sinistri
                          Lnn. lumbales
                          Lnn. mediastinales posteriores
                          Lnn. pylorici
                          Lnn. splenici
                          Lnn. hepatici

Spleen

The spleen possesses a subserous lymphatic network. These superficial lymphatics and the deep lymph vessels which accompany the blood vessels terminate in lymph nodes of the hilus.

Regional lymph nodes   *Lnn. splenici*

Pancreas (Fig. 41)

The rich lymphatic network of the pancreas communicates with that of the duodenum. The lymph vessels themselves may be found in the interlobular spaces. At the pancreatic surface they accompany the blood vessels and are drained by collecting lymph vessels which may be attributed to the head, neck, body or tail of the organ [4].

At the anterior surface of the head, the upper collecting lymph vessels drain via the superior pancreaticoduodenal nodes to the common hepatic artery group. The middle and lower pathways are associated with the superior mesenteric artery lymph nodes. These three pathways terminate in a node between the celiac artery and the superior mesenteric artery (Ln. celiacomesentericus dexter superficialis according to Deki and Sato [8]). The lymph vessels of the inferior third of the retrovenous pancreatic segment (uncinate process) terminate in the left subrenal interaorticocaval lymph node [34].

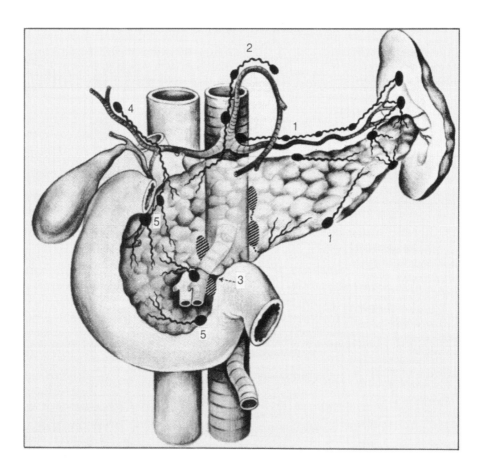

Fig. 41. Lymph node regions of the pancreas. (Adapted from [11])

1 Pancreatic lymph nodes   4 Hepatic lymph nodes
2 Celiac lymph nodes   5 Superior and inferior pancreatico-
3 Mesenteric lymph nodes   duodenal lymph nodes

The lymph of the posterior surface of the head is drained to posterior pancreatico-duodenal nodes and an interaorticocaval node.

The lymph vessels of the anterior surface of the neck of the pancreas join the vessels of the head, thus emptying to the common hepatic chain, whereas the posterior surface drains to lymph nodes between the celiac artery and the superior mesenteric artery (Lnn. celiacomesenterici dextri superficiales et profundi according to Deki and Sato [8]).

The anterior surface of the median subsegment of the body of the pancreas is drained superiorly to common hepatic nodes or nodes around the superior mesenteric artery. The upper portion of the left subsegment of the body is associated with the splenic lymph chain, whereas lymph vessels of the lower portion accompany the inferior pancreatic artery, thus terminating in a node situated to the left of the celiac artery and the superior mesenteric artery (Ln. celiacomesentericus sinister according to Deki and Sato [8]).

Lymph from the posterior surface of the median subsegment of the pancreatic body flows to common hepatic, left gastric, splenic and superior mesenteric nodes terminating in the lymph nodes between the celiac artery and the superior mesenteric artery.

The majority of the lymph vessels of the tail of the pancreas drain via the splenic route with the regional lymph nodes at the hilus.

In summary, almost all lymph vessels of the right half of the pancreas terminate in the right deep celiacomesenteric node. The efferent vessels from this node empty in the interaorticocaval nodes which may be found in the angles formed by the inferior vena cava and the left renal vein. On the other hand, the lymph from the left half of the pancreas is collected to the left celiacomesenteric lymph node which is located around the axis formed by the celiac trunk and the superior mesenteric artery. From this node the efferent lymph vessels drain to the left latero-aortic lymph nodes lying above and below the left renal vein.

Large efferent vessels start both from the interaorticocaval nodes and the lateroaortic nodes toward the retro-aortic space where they form an ascending lymphatic system together with the vessels from the lower abdomino-aortic nodes [8].

Regional lymph nodes	*Lnn. pancreatici (superiores et inferiores)*
	Lnn. splenici
	Lnn. pancreaticoduodenales
	Lnn. coeliaci
	Lnn. mesenterici
	(Lnn. mesenterici superiores)
	Lnn. lumbales sinistri
	Lnn. lumbales intermedii

Small Bowel

The small bowel has a mucosal, submucosal, intramuscular and subserous lymphatic network. The lymph of the superior and descending parts of the duodenum is collected by the superior pancreaticoduodenal lymph nodes and the subpyloric nodes. From there, the lymph flow is directed to the hepatic nodes and the celiac nodes. The horizontal and ascending parts of the duodenum are drained via the inferior pancreaticoduodenal lymph nodes to nodes around the superior mesenteric artery.

The regional lymph nodes of the jejunum and the ileum are the very numerous mesenteric nodes. The terminal part of the ileum, i.e., the distal 15 cm, is drained to the ileocolic chain.

Regional lymph nodes   *Lnn. pancreaticoduodenales superiores et inferiores*
    Lnn. pylorici (Lnn. subpylorici)
    Lnn. mesenterici
    Lnn. ileocolici

Colon (Fig. 42)

From the mucosa the colonic lymphatics pass into the submucosa, where they follow the blood vessels. They then penetrate the muscularis mucosae, join the subserosal network and form collecting lymph vessels in the mesentery, which again accompany the blood vessels. The first, most important and most numerous regional lymph nodes are the paracolic nodes. Via the intermediate nodes the lymph flows to nodes at the base of the mesentery and from there to inferior and superior mesenteric lymph nodes. The efferent vessels of the inferior mesenteric nodes terminate in superior mesenteric or latero-aortic nodes or in the left lumbar trunk. The lymph of the superior mesenteric nodes drains to several intestinal trunks, to the left lumbar trunk or to vessels which empty directly into the cisterna chyli.

In detail, the paracolic lymph nodes of the ascending colon drain the lymph along the right colic arcades to the ileocolic chain or directly to intermediate nodes along the right colic vessels or to central nodes anterolateral to the junction of the right colic vein and the superior mesenteric vein. The cecum possesses lymphatic vessels which empty to pre- and retrocecal nodes and the ileocolic chain. The subserous network of the vermiform appendix gives rise to four or five collecting lymph vessels which accompany the appendicular artery and terminate in the appendicular nodes. In addition, some appendicular lymph is drained to the regional lymph nodes of the cecum. Lymph from the proximal transverse colon flows to paracolic nodes or directly to intermediate nodes along the middle colic artery or to central nodes at the origin of the middle colic artery from the superior mesenteric artery. Similarly, the lymph vessels of the distal two-thirds of the transverse colon may proceed directly to intermediate nodes along the left colic artery. The lymph of the sigmoid is drained to the sigmoid lymph nodes and from there to pre- and latero-aortic nodes.

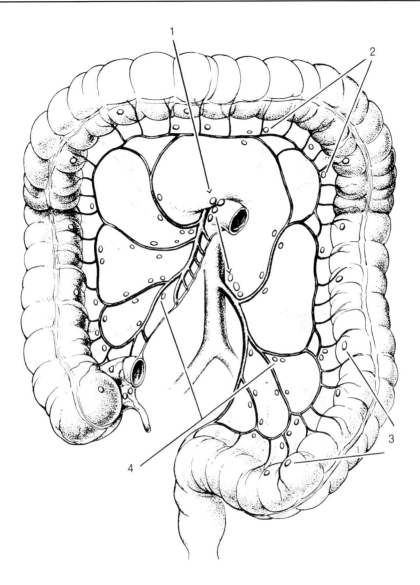

Fig. 42. Topography of the regional lymph nodes of the colon. (Adapted from [15])

1 Superior and inferior mesenteric lymph nodes
2 Mesocolic lymph nodes
3 Paracolic lymph nodes
4 Intermediate lymph nodes

Like the avascular area of Treves in the ileocecal region, the avascular mesentery at the splenic flexure provides a natural hiatus of lymphatic drainage: proximally the lymph is collected to superior mesenteric nodes, distally via left colic and inferior mesenteric nodes to pre-aortic nodes. As for the descending colon, no direct communications to central nodes have been demonstrated.

Regional lymph nodes	*Lnn. mesocolici*
	(Lnn. paracolici, Lnn. colici dextri, medii et sinistri)
	Lnn. ileocolici
	Lnn. pre- et retrocecales
	Lnn. appendiculares
	Lnn. mesenterici
	(Lnn. mesenterici superiores)
	Lnn. mesenterici inferiores
	(Lnn. sigmoidei)

Suprarenal Gland

The lymph vessels of the cortex accompany blood vessels penetrating through the capsule to join with the lymph vessels of the capsule. The cortical lymph vessels communicate with the lymph vessels of the medulla, which may be found in close relationship to the suprarenal veins.

The collecting lymph vessels, which accompany the superior suprarenal artery terminate in lymph nodes around the celiac artery. The latero-aortic lymph nodes represent the regional lymph nodes for collecting lymph vessels which accompany the middle suprarenal artery and the suprarenal vein. In addition, some lymph vessels may have connections to lymphatics of the liver, while others may penetrate through the diaphragm to terminate in lymph nodes of the posterior mediastinum.

Regional lymph nodes	*Lnn. coeliaci*
	Lnn. lumbales sinistri
	Lnn. lumbales intermedii
	Lnn. mediastinales posteriores

Kidney

The adipose tissue of the renal capsule possesses a superficial and a deep lymphatic network giving rise to collecting lymph vessels that accompany the renal artery and vein. The lymph is drained to pre- or latero-aortic lymph nodes. The lymphatics of the adipose capsule communicate with the lymphatic networks of neighboring organs, e.g., liver, diaphragm, cecum, appendix and colon, but there seem to be no connections to the lymph vessels of the fibrous capsule or the parenchyma of the kidney.

According to Rawson [38], two lymphatic systems may be separated in the parenchyma. The first starts at Bowman's capsule, enlarges by forming nets around the cortical blood vessels which accompany the interlobular vessels, follows the arcuate and interlobar vessels and leaves the kidney at the hilus. The second system begins just beneath the mucosa of the papillae, accompanies the medullar blood vessels and forms lymph channels surrounding the arcuate arteries and veins.

According to Rouvière [42], three main collecting lymph vessels may be described at the renal pedicle. They lie in front of, between and behind the renal vessels: The posterior collecting lymph vessels, three to five in number, terminate on the right side in interaorticocaval lymph nodes and on the left in latero-aortic nodes between the renal artery and the inferior mesenteric artery. The middle collecting lymph vessels, one or two in number, drain on the right side to a latero-aortic node, whereas on the left side they may empty either into a latero-aortic node or into a node near the junction of the suprarenal and renal veins. On the right side, the anterior collecting lymph vessels, three or four in number, terminate in the latero-aortic lymph nodes and occasionally in a precaval lymph node and a node lying below the termination of the renal vein. On the left side, these vessels drain to the latero-aortic nodes, a node at the junction of the left suprarenal and renal veins and a node at the termination of the left spermatic vein into the renal vein.

The latero-aortic lymph nodes collect the lymph of the renal pelvis.

Regional lymph nodes   *Lnn. lumbales sinistri*
                        Lnn. lumbales intermedi
                        Lnn. lumbales dextri

Ureter

The lymphatic system of the ureter may be divided into three sections as follows: The superior segment extends from the origin to the intersection with the spermatic or ovarian vessels. The lymph of this segment is drained to latero-aortic nodes and irregularly to iliac nodes. The middle segment extends from the intersection with the spermatic or ovarian vessels down to that with the iliac vessels. Its lymph empties to a latero-aortic node close to the origin of the inferior mesenteric artery. The lymph of the inferior segment, i. e., the iliac and pelvic part of the ureter, is collected by iliac lymph nodes.

Regional lymph nodes   *Lnn. lumbales sinistri*
                        Lnn. lumbales intermedii
                        Lnn. lumbales dextri
                        Lnn. iliaci communes
                        Lnn. iliaci externi

Topographical Anatomy
of Regional Lymph Node Groups of Abdomen

Parietal and visceral lymph node groups of the abdomen have to be differentiated.

Parietal Nodes (Figs. 43–50)

Lnn. Lumbales Sinistri (Figs. 46–50)

These lymph nodes may be found adjacent to the abdominal aorta. They receive lymph directly from the adrenal gland, kidney, ureter, testis, ovary, fallopian tube, uterine fundus, abdominal wall and iliac lymph nodes. The efferent lymph vessels of these nodes terminate in the lumbar trunk. Three groups may be separated.

Lnn. aortici laterales: to the left of the aorta.

Lnn. pre-aortici: anterior to the aorta.

Lnn. postaortici: embedded between the aorta and the spine.

Lnn. Lumbales Intermedii (Figs. 47–49)

These lymph nodes drain the lymph from the same organs as the left lumbar nodes. They lie between the aorta and the inferior vena cava.

Lnn. Lumbales Dextri (Figs. 45–50)

These have the same function as the left lumbar and intermediate groups and may be divided into three subgroups.

Lnn. cavales laterales (Fig. 46): just to the right of the inferior vena cava.

Lnn. precavales (Fig. 46): anterior to the vena cava.

Lnn. postcavales (Figs. 46–48, 58): posterior to the vena cava.

Lnn. Phrenici Inferiores (Figs. 43, 44)

These nodes are located on the inferior surface of the diaphragm near the aortic hiatus.

Lnn. Epigastrici Inferiores

These may be demonstrated around the inferior epigastric vessels.

Visceral Nodes (Figs. 43–50, 57, 58, 64)

Lnn. Coeliaci

These lymph nodes are grouped around the celiac artery and serve as the secondary draining nodes for the stomach, duodenum, liver, gallbladder, pancreas and spleen. The efferent vessels terminate in the intestinal trunk or the cisterna chyli.

Lnn. Hepatici (Fig. 45)

These are located near the hepatic artery in the hepatic hilus and along the hepatoduodenal ligament and carry lymph mainly from the liver to the celiac nodes. Two single nodes belonging to the hepatic nodes can be found within the biliary system.

Nodus cysticus (Fig. 45): at the neck of the gallbladder.

Nodus foraminalis: close to the epiploic foramen (foramen Winslowi) near the cystic duct and the posterior surface of the duodenum.

Lnn. Gastrici (Dextri et Sinistri) (Figs. 43, 44)

The course of this chain follows the right and the left gastric artery along the lesser curvature. It drains to the celiac nodes.
  The cardia sometimes exhibits a lymphatic ring, the annulus lymphaticus cardiae.

Lnn. Gastro-omentales (Dextri et Sinistri) (Figs. 43–45)

This group accompanies the gastroepiploic vessels along the greater curvature and drains either to hepatic, pancreatic or splenic nodes.

Lnn. Pylorici

Draining to hepatic and celiac nodes, these nodes can be found around the pylorus and may be subdivided into three groups.

Ln. suprapyloricus: a node on the upper surface of the pylorus in proximity to the left gastric artery.

Lnn. subpylorici: under the pyloric canal in close proximity to the right gastroepiploic artery.

Lnn. retropylorici: behind the pylorus near the gastroduodenal artery.

Lnn. Pancreaticoduodenales (Superiores et Inferiores)
(Figs. 46, 48)

These small lymph nodes are situated between the head of the pancreas and the duodenum along the anterior and the posterior branch of the pancreaticoduodenal artery. Their efferent vessels terminate in the hepatic and mesenteric nodes.

Lnn. Pancreatici (Superiores et Inferiores) (Figs. 45–47)

This group may be divided into a superior and an inferior chain along the upper and lower border of the organ respectively. The efferent vessels empty into mesenteric, pancreaticoduodenal and splenic nodes.

Lnn. Splenici (Fig. 45)

These are situated in the splenic hilus and drain to celiac nodes.

Lnn. Mesenterici

This, the largest lymph node group in the body, consists of two principal subgroups which drain mainly to the celiac nodes:

Lnn. juxtaintestinales (Figs. 45–50, 57, 58, 64): in close relation to the most peripheral vascular arcades.

Lnn. Superiores (Figs. 46–49): grouped along the superior mesenteric artery.

Lnn. Appendiculares

This inconstant group lies adjacent to the appendiculars artery and drains to central mesenteric nodes.

Lnn. Pre- et Retrocecales (Fig. 57)

These accompany the anterior and posterior cecal artery and also drain to central mesenteric nodes.

Lnn. Ileocolici

This chain lies along the ileocolic artery and discharges its lymph into celiac nodes.

Lnn. Mesocolici (Figs. 44, 48, 49, 50)

These nodes drain most parts of the colon and accompany the colic arteries. The efferent vessels drain to the central mesenteric nodes or, like left colic and sigmoid nodes, to the inferior mesenteric nodes. The group consists of the following subgroups.

Lnn. paracolici (Figs. 46, 47, 49, 57, 58): along the marginal artery.

Lnn. colici dextri, medii et sinistri: accompanying the right, middle and left colic arteries respectively.

Lnn. Mesenterici Inferiores

This group is situated from the origin of the inferior mesenteric artery down to the branching of the left colic artery. It receives lymph from the descending colon, the sigmoid and the rectum and drains to pre- and latero-aortic nodes.

Lnn. sigmoidei (Fig. 58): closely attached to the branches of the sigmoid artery.

Lnn. rectales superiores: as these are regional nodes for the rectum, their topographical anatomy will be described in the chapter on the pelvic lymph nodes.

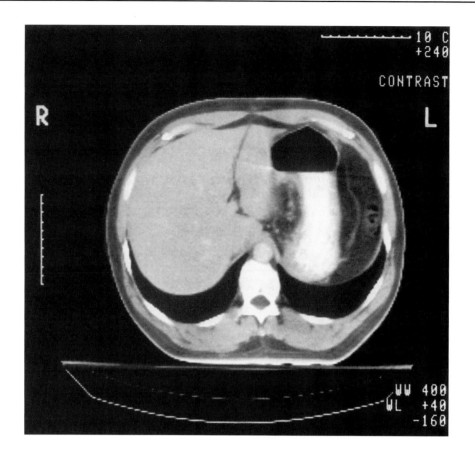

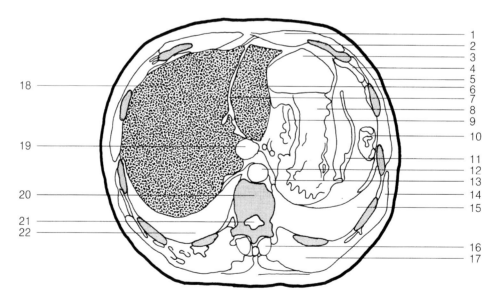

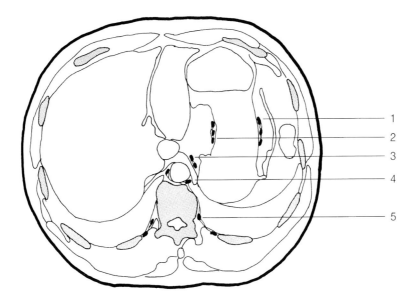

Fig. 43. Axial section of the abdomen   ▲

◀ 1 Rectus abdominis muscle
   2 Rib
   3 Stomach
   4 Diaphragm
   5 Intercostal muscles
   6 Liver (left lobe)
   7 Falciform ligament of liver
   8 Stomach
   9 Right gastric artery
10 Colon
11 Gastrosplenic ligament
12 Descending aorta
13 Diaphragm
14 Latissimus dorsi muscle
15 Lung
16 Transversospinalis muscle
17 Erector muscle of spine
18 Liver (right lobe)
19 Inferior vena cava
20 Vertebra
21 Vertebral canal
22 Lung

1 Gastro-omental lymph nodes
2 Right gastric lymph nodes
3 Inferior phrenic lymph nodes
4 Superior phrenic lymph nodes
5 Intercostal lymph nodes

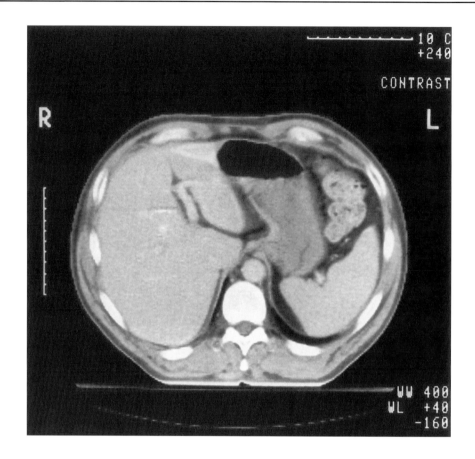

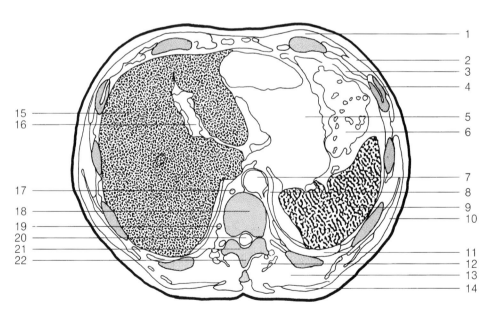

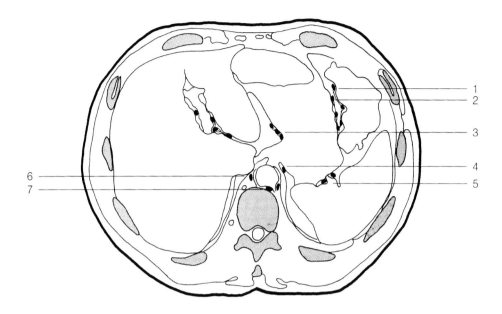

Fig. 44. Axial section of the abdomen ▲

◄ 1 Rectus abdominis muscle
  2 Intercostal muscles
  3 Diaphragm
  4 Rib
  5 Stomach
  6 Colon
  7 Descending aorta
  8 Diaphragm
  9 Spleen
10 Latissimus dorsi muscle
11 Costodiaphragmatic recess of pleura
12 Transversospinalis muscle
13 Longissimus muscle
14 Trapezius muscle
15 Liver
16 Branch of portal vein
17 Azygos vein
18 Vertebra
19 Ascending lumbar vein
20 Vertebral canal
21 Vertebral arch
22 Spine of a vertebra

1 Mesocolic lymph nodes
2 Gastro-omental lymph nodes
3 Gastric lymph nodes
4 Inferior phrenic lymph nodes
5 Splenic lymph nodes
6 Superior phrenic lymph nodes
7 Prevertebral lymph nodes

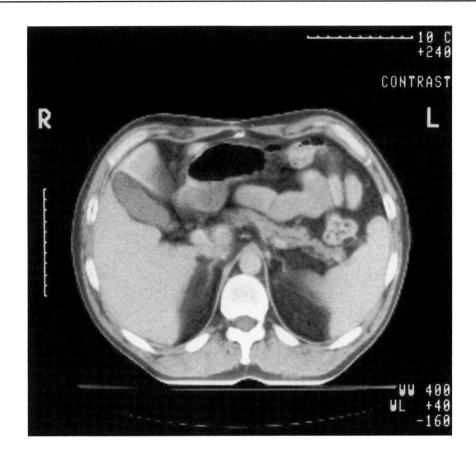

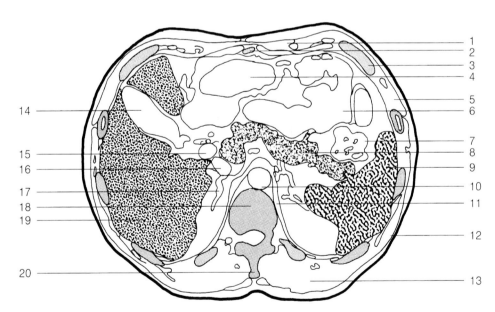

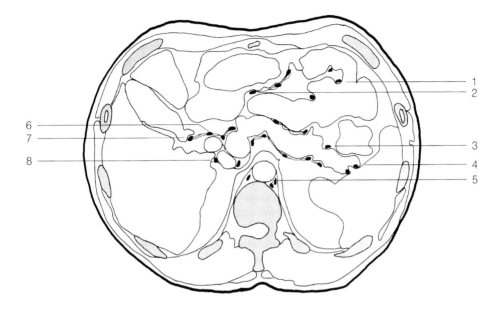

Fig. 45. Axial section of the abdomen   ▲

◀ 1 Rectus abdominis muscle
   2 Xiphoid process
   3 Rib
   4 Stomach
   5 Intercostal muscles
   6 Jejunum
   7 Colon
   8 Pancreas
   9 Spleen
 10 Descending aorta
 11 Diaphragm
 12 Latissimus dorsi muscle
 13 Erector muscle of spine
 14 Gall bladder
 15 Portal vein
 16 Inferior vena cava
 17 Suprarenal gland
 18 Vertebra
 19 Liver
 20 Spine of a vertebra

 1 Juxtaintestinal lymph nodes
 2 Gastro-omental lymph nodes
 3 Pancreatic lymph nodes
 4 Splenic lymph nodes
 5 Superior phrenic lymph nodes
 6 Hepatic lymph nodes
 7 Cystic node
 8 Right lumbar lymph nodes

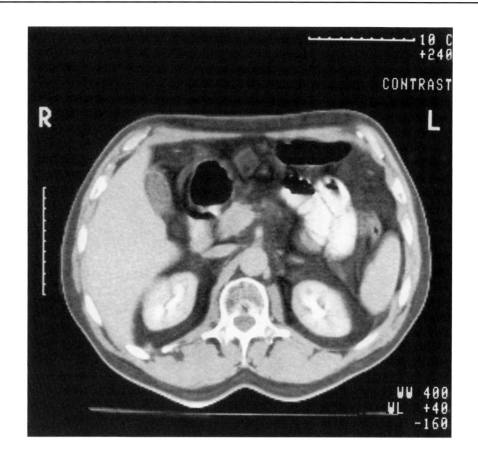

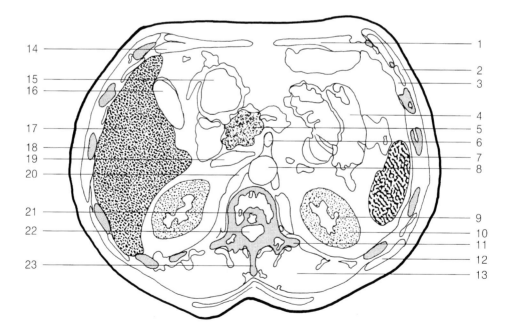

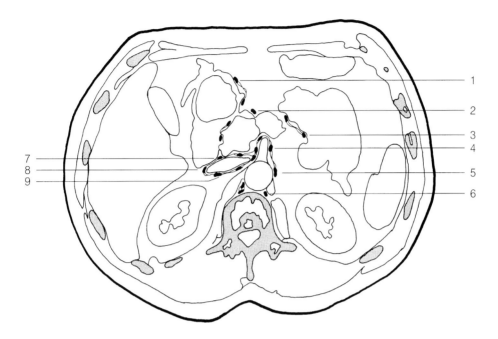

Fig. 46. Axial section of the abdomen ▲

◀ 1 Rectus abdominis muscle
2 External oblique muscle
3 Intercostal muscles
4 Jejunum
5 Pancreas
6 Superior mesenteric artery
7 Spleen
8 Descending aorta
9 Renal calices
10 Kidney
11 Transverse process
12 Latissimus dorsi muscle
13 Erector muscle of spine
14 Transverse muscle of abdomen
15 Colon
16 Gall bladder
17 Liver
18 Rib
19 Inferior vena cava
20 Diaphragm
21 Vertebra
22 Vertebral canal
23 Spine of a vertebra

1 Paracolic lymph nodes
2 Pancreatic lymph nodes
3 Juxtaintestinal lymph nodes
4 Superior mesenteric lymph nodes
5 Left lumbar lymph nodes
6 Superior phrenic lymph nodes
7 Precaval lymph nodes (right lumbar lymph nodes)
8 Laterocaval lymph nodes (right lumbar lymph nodes)
9 Retrocaval lymph nodes (right lumbar lymph nodes)

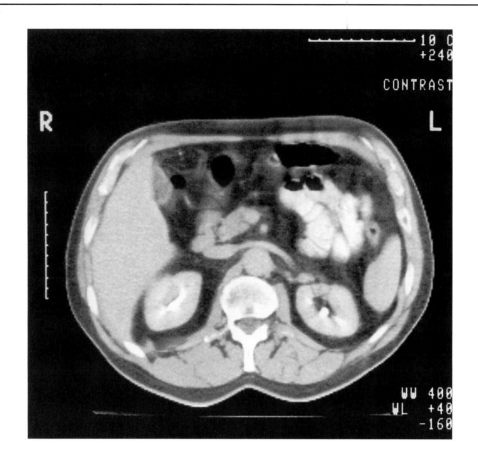

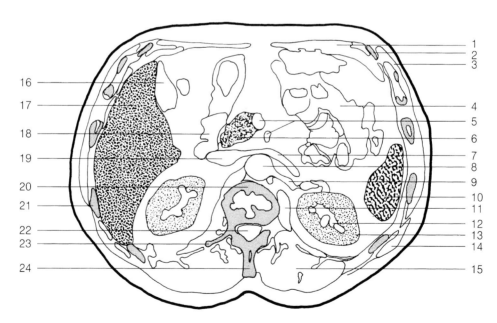

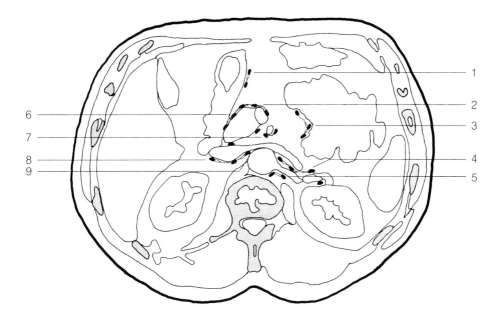

Fig. 47. Axial section of the abdomen ▲

◄ 1 Rectus abdominis muscle
   2 Vertebral arch
   3 External oblique muscle
   4 Jejunum
   5 Superior mesenteric artery and vein
   6 Body of pancreas
   7 Renal vein
   8 Abdominal aorta
   9 Renal artery
  10 Spleen
  11 Greater psoas muscle
  12 Renal calices
  13 Kidney
  14 Latissimus dorsi muscle
  15 Erector muscle of spine
  16 Gall bladder
  17 Liver
  18 Descending part of duodenum
  19 Inferior vena cava
  20 Vertebra
  21 Rib
  22 Vertebral canal
  23 Transverse process
  24 Spine of a vertebra

1 Paracolic lymph nodes
2 Pancreatic lymph nodes
3 Juxtaintestinal lymph nodes
4 Left lumbar lymph nodes
5 Retrocaval lymph nodes
6 Pancreaticoduodenal lymph nodes
7 Superior mesenteric lymph nodes
8 Right lumbar (pre- and retrocaval, laterocaval) lymph nodes
9 Intermediate lumbar lymph nodes

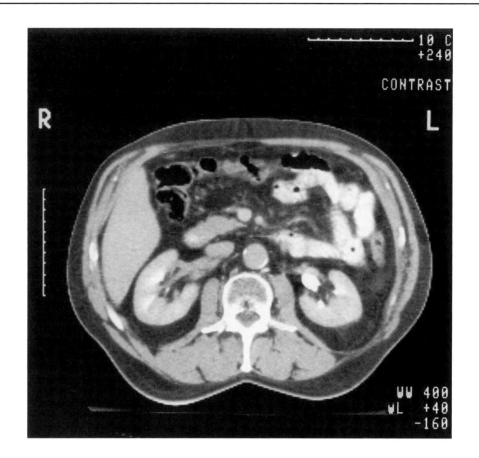

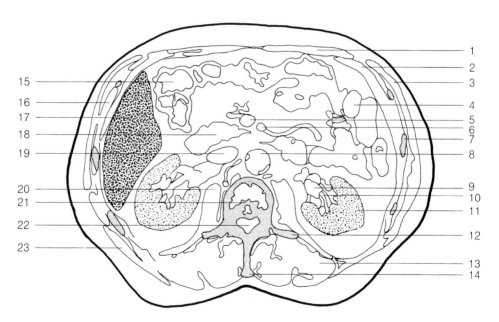

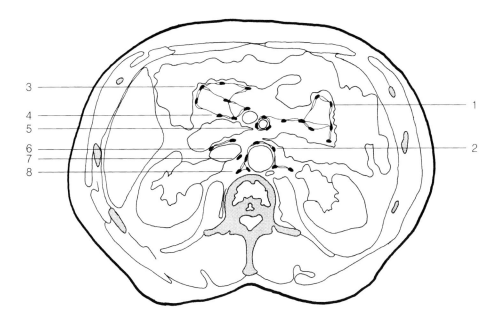

Fig. 48. Axial section of the abdomen ▲

◄ 1 Rectus abdominis muscle
   2 Transverse muscle of abdomen
   3 External oblique muscle
   4 Jejunum
   5 Superior mesenteric vein
   6 Superior mesenteric artery
   7 Rib
   8 Renal pelvis
   9 Sinus of kidney
  10 Renal artery
  11 Kidney
  12 Transverse process
  13 Erector muscle of spine
  14 Spine of a vertebra
  15 Colon
  16 External oblique muscle
  17 Liver
  18 Horizontal part of duodenum
  19 Superior vena cava
  20 Greater psoas muscle
  21 Vertebra
  22 Vertebral canal
  23 Latissimus dorsi muscle

   1 Juxtaintestinal lymph nodes
   2 Left lumbar lymph nodes
   3 Mesocolic lymph nodes
   4 Inferior pancreaticoduodenal lymph nodes
   5 Superior mesenteric lymph nodes
   6 Right lumbar lymph nodes
   7 Intermediate lumbar lymph nodes
   8 Retrocaval lymph nodes

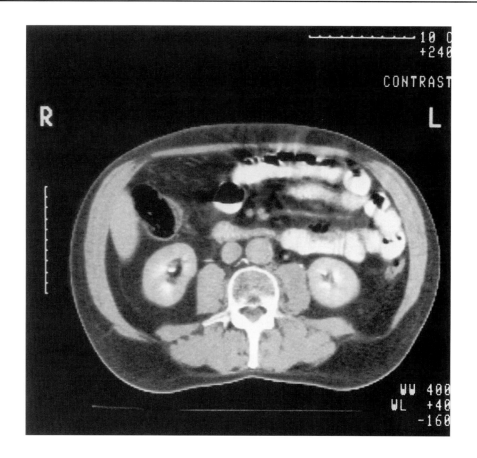

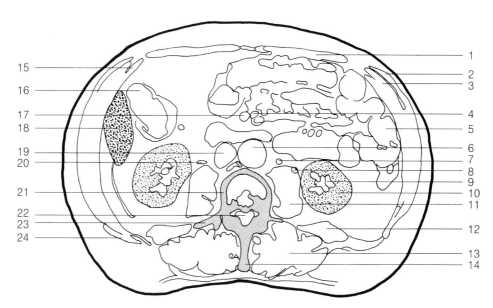

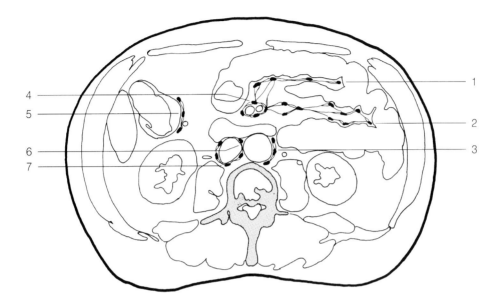

Fig. 49. Axial section of the abdomen   ▲

◄ 1 Rectus abdominis muscle
   2 External oblique muscle
   3 Internal oblique muscle
   4 Superior mesenteric artery
   5 Jejunum
   6 Abdominal aorta
   7 Inferior mesenteric vein
   8 Ureter
   9 Renal calices
  10 Kidney
  11 Greater psoas muscle
  12 Quadratus lumborum muscle
  13 Erector muscle of spine
  14 Spine of a vertebra
  15 Transverse muscle of abdomen
  16 Colon
  17 Inferior mesenteric vein
  18 Liver
  19 Inferior vena cava
  20 Ureter
  21 Vertebra
  22 Vertebral canal
  23 Transverse process
  24 Latissimus dorsi muscle

1, 2 Juxtaintestinal lymph nodes
   3 Left lumbar lymph nodes
   4 Superior mesenteric lymph nodes
   5 Paracolic lymph nodes (mesocolic lymph nodes)
   6 Intermediate lumbar lymph nodes
   7 Caval lymph nodes (right lumbar lymph nodes)

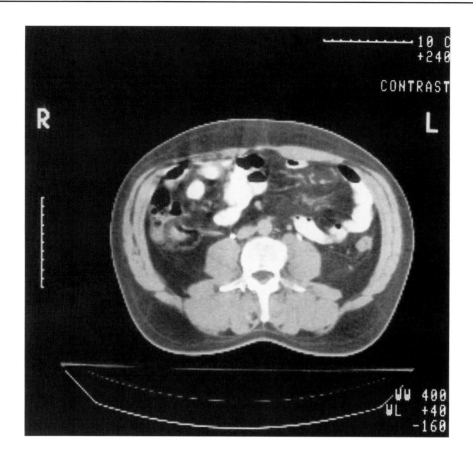

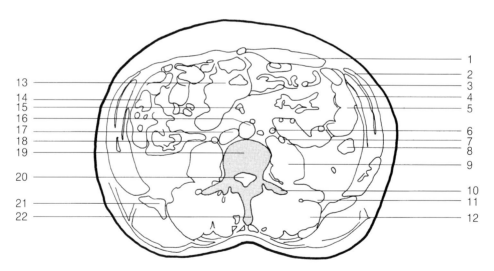

Fig. 50. Axial section of the abdomen   ▲

◄ 1 Rectus abdominis muscle
   2 External oblique muscle
   3 Internal oblique muscle
   4 Transverse muscle of abdomen
   5 Jejunum
   6 Inferior mesenteric vein
   7 Ureter
   8 Colon
   9 Greater psoas muscle
  10 Transverse process
  11 Quadratus lumborum muscle
  12 Erector muscle of spine
  13 Ileum
  14 Colon
  15 Ileal veins
  16 Abdominal aorta
  17 Inferior vena cava
  18 Ureter
  19 Vertebra
  20 Vertebral canal
  21 Latissimus dorsi muscle
  22 Spine of a vertebra

1, 2 Juxtaintestinal lymph nodes
  3 Left lumbar lymph nodes
4, 5 Mesocolic lymph nodes
  6 Juxtaintestinal lymph nodes
  7 Right lumbar lymph nodes

Pelvis and Inguinal Region

General Considerations

The lymph vessels of the pelvic organs drain either to visceral nodes in the subperitoneal adipose tissue such as the paravesical, parauterine, paravaginal and pararectal nodes or directly to the regional nodes.

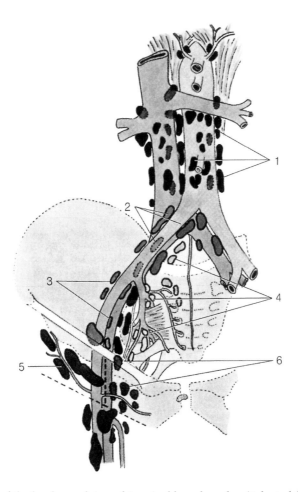

Fig. 51. Topography of the lumbar, pelvic and inguinal lymph nodes. (Adapted from [24])

1 Lumbar lymph nodes
2 Common iliac lymph nodes
3 External iliac lymph nodes

4 Internal iliac lymph nodes
5 Superficial inguinal lymph nodes
6 Deep inguinal lymph nodes

Like the lumbar lymph nodes, the iliac nodes (Fig. 51) accompany the blood vessels and consist of three different chains: medial, intermediate and lateral. For the external iliac nodes, the medial chain is the main channel, for the common iliac nodes, the lateral chain. Connections in both directions between the external and the internal iliac nodes have been demonstrated. A "segmental passing over" of single iliac nodes may be present [20].

Lymph from the bladder, the prostate and the uterine cervix is drained mainly to the external iliac nodes, partly to common iliac nodes. The lymph vessels of the upper part of the vagina terminate in external and common iliac nodes, whereas the internal iliac nodes represent the collecting nodes for the lower part of the vagina. The pelvic portion of the rectum drains to the internal and common iliac nodes. The regional nodes for the uterine corpus, the fallopian tube, the ovary and the testis are the lumbar nodes. This lymphatic pathway represents a relict of the descent of the fetal gonads.

The inguinal nodes are grouped around the termination of the great saphenous vein and along the inguinal ligament. In addition to their function draining lymph from the lower extremity, they receive lymph from pelvic organs such as the external genitals and the lower third of the vagina.

Lymphatic Drainage Regions

Rectum and Anal Canal (Fig. 52)

The pelvic part of the rectum and the anal canal represent different lymph drainage territories. The pectinate line separates the two territories; however, communications between them do exist, i. e., the lymphatics of the anal canal communicate with the perineal and rectal lymphatics. The lymph of the perineum and the anal canal is drained to the superficial inguinal lymph nodes, whereas the lymph of the pelvic territory flows to iliac and lumbar nodes [21].

The rectal lymph is either collected by the pararectal nodes or the superior rectal nodes or directly drained by three main collecting lymph vessels (upper, middle, lower). The pararectal lymph nodes are situated above the levator muscles on the posterolateral surface of the rectum between the rectal fascia and the muscular coat and in the angles formed by the branches of the superior rectal artery. The superior rectal nodes may be found in the mesorectum along the superior rectal artery. The lowest node of this group, the so-called principal node of the rectum, lies at the level of the bifurcation of the superior rectal artery. The lymphatic drainage above the level of the middle rectal valve, which coincides with the level of the anterior peritoneal reflection 7–8 cm above the anal verge, is performed by the upper collecting lymph vessel. This vessel accompanies the branches of the superior rectal artery and drains via the inferior mesenteric and sigmoid nodes to the latero-aortic and preaortic nodes. The middle collecting lymph vessel partly follows the middle rectal or the internal pudendal artery and drains to the internal iliac chain, and partly accompanies the middle and lateral sacral artery, dis-

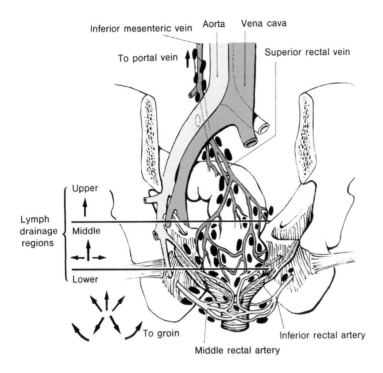

Fig. 52. The lymphatic drainage of the rectum. The arrows indicate the main lymph flow directions. (Adapted from [14])

charging into the lateral sacral and promontorial nodes. The lower collecting lymph vessel, which drains the distal rectum, the columns of Morgagni, the anal canal and the perianal skin, usually either finds its way to the superficial inguinal nodes or accompanies the inferior hemorrhoidal vessels through the ischiorectal fascia to the internal pudendal artery and terminates in the internal iliac nodes.

Regional lymph nodes   *Lnn. pararectales*
                            Lnn. mesenterici inferiores
                            (Lnn. rectales superiores)
                            Lnn. iliaci interni
                            Lnn. inguinales superficiales

Urinary Bladder (Fig. 53)

A rich lymphatic network can be demonstrated in the mucosa and the muscular layers of the bladder. These lymphatics communicate with a network on the surface of the bladder, the latter giving rise to collecting lymph vessels which unite in the prevesical space and terminate particularly in the middle and cranial nodes of the medial and intermediate external iliac groups. The efferent lymph vessels sometimes are interrupted by paravesical nodes.

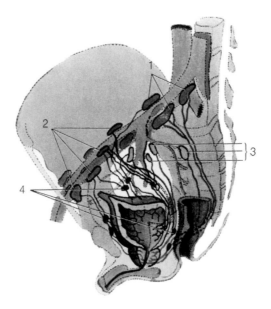

Fig. 53. The lymphatic drainage of the urinary bladder and the prostate. (Adapted from [24])

1 Common iliac lymph nodes   3 Internal iliac lymph nodes
2 External iliac lymph nodes   4 Paravesicular lymph nodes

In detail, the anterior wall of the bladder is drained by collecting lymph vessels situated at the middle third of the lateral vesical border. These vessels descend toward the origins of the umbilical artery and the superior vesical artery, unite with the collecting lymph vessels of the posterior wall and terminate in the external iliac nodes [42].

The collecting lymph vessels of the posterior wall either run to the posterolateral angle of the bladder, cross the umbilical artery and drain to the medial and intermediate external iliac chain, or empty into the collecting lymph vessels of the trigone. Less frequently, the lymph is drained to the medial lacunar node or internal iliac nodes. This also applies to the lymph of the trigone, but the majority of the collecting lymph vessels of the trigone accompany the uterine or the deferential artery and terminate in the medial and intermediate groups of external iliac nodes.

Regional lymph nodes   *Lnn. paravesiculares*
                            Lnn. iliaci externi
                            (Lnn. iliaci exteni mediales et intermedii,
                            Ln. lacunaris mediales)
                            Lnn. iliaci interni

Urethra

In the male, the lymph vessels of the spongiose part of the urethra drain to the regional lymph nodes of the corpora cavernosa. The lymphatics of the prostatic part share their regional lymph nodes with the prostate. As for the membranous part of the urethra, three main collecting lymph vessels may be distinguished: The first one accompanies the internal pudendal artery and drains into the inferior gluteal nodes. The second runs behind the symphysis and empties into the medial lacunar node. The third is situated in front of the prostate and unites with the efferent lymph vessels of the bladder which terminate in the middle and upper nodes of the medial external iliac chain.

The regional lymph nodes of the female urethra correspond to those of the prostatic and membranous parts of the male urethra.

Regional lymph nodes   *Lnn. iliaci externi*
                           (Lnn. iliaci externi mediales,
                           Ln. lacunaris medialis)
                           Lnn. iliaci interni
                           (Lnn. gluteales inferiores)
                           Lnn. inguinales superficiales

Male Internal Genitals

Prostate (Fig. 54)

The prostatic lymph vessels arise from the glandular acini, find their way through the capsule and there form a rich lymphatic network, especially on the posterior and superior surface.

The collecting lymph vessels of the anterior lobe drain mainly to the intermediate and medial external iliac chain, either directly or via intermediate and prevesical nodes. Another lymphatic pathway accompanies the internal pudendal artery, terminating in the inferior gluteal nodes.

The lymphatics of the posterior lobe of the prostate give rise to three collecting lymph vessels: the first starts from the base of the gland, follows the seminal vesicle, passes above the terminal segment of the ureter and finally drains to the middle and upper nodes of the intermediate external iliac chain. The second arises from the inferior and posterior part of the prostate, accompanies the prostatic artery and empties into the internal iliac nodes. The third travels dorsally to superior gluteal lymph nodes and promontory lymph nodes along the sacroprostatic ligament [21].

The prostatic lymphatic network richly anastomoses with the lymphatics of the rectum, bladder, ampulla of the ductus deferens and seminal vesicles.

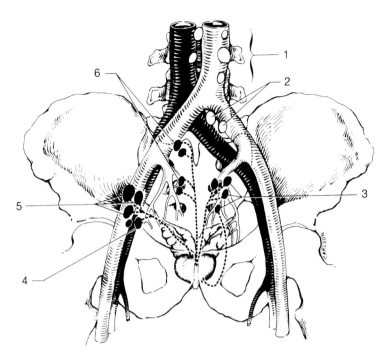

Fig. 54. The regional lymph nodes of the prostate. (Adapted from [13])

1 Periaortic nodes   4 "Obturator node"
2 Common iliac nodes   5 External iliac nodes
3 Internal iliac nodes   6 Presacral nodes

Regional lymph nodes   *Lnn. iliaci externi*
                          (Lnn. iliaci externi mediales et intermedii)
                          Lnn. iliaci interni
                          (Lnn. gluteales superiores et inferiores)
                          Lnn. iliaci communes
                          (Lnn. iliaci communes promontorii)
                          Lnn. paravesiculares

Ductus Deferens and Seminal Vesicles

The lymph vessels of the proximal part of the ductus deferens ascend together with collecting lymph vessels of the testis and epididymis to the lumbar nodes. The lymph vessels of the middle part terminate either in the intermediate and medial nodes of the external iliac chain or in the lateral lacunar node. Together with the efferent lymph vessels of the seminal vesicle, the collecting lymph vessels of the terminal part of the ductus deferens join the efferent vessels of the urinary bladder and the prostate and empty into medial and intermediate nodes of the external iliac chain and internal iliac nodes [21].

Regional lymph nodes   *Lnn. iliaci externi*
                              (Lnn. iliaci externi mediales et intermedii)
                              Lnn. iliaci interni
                              Lnn. lumbales dextri et sinistri

Testicle and Epididymis (Fig. 55)

The testicle and the epididymis possess a common lymphatic drainage system which consists of four to eight collecting lymph vessels. These vessels accompany the spermatic vessels, crossing the ureter medially and then spreading out to the lumbar lymph nodes from the renal vein to the aortic bifurcation. On the right side, the lymph vessels drain to the laterocaval and precaval lymph nodes, on the left side, to the latero-aortic,

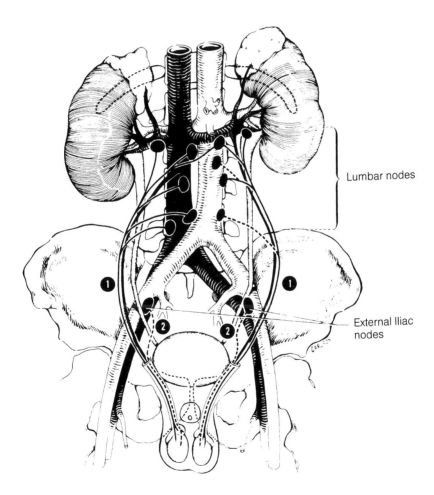

Fig. 55. The lymphatic drainage of the testis and epididymis. Note that on the right, most lymphatic channels (1) drain to lymph nodes between the renal vein and the aortic bifurcation, whereas on the left, most lymphatic channels (1) drain to nodes located below the renal vein. On both sides, some lymphatic channels terminate in the external iliac chain (2). (Adapted from [13])

pre-aortic and intermediate lumbar nodes. Sometimes the collecting lymph vessels cross the midline [21].

Some of the lymph vessels leave the spermatic vessels after reaching the vesical peritoneum, turn laterally and empty into intermediate nodes of the external iliac chain.

Regional lymph nodes   *Lnn. lumbales dextri et sinistri*
                            Lnn. iliaci externi
                            (Lnn. iliaci externi intermedii)

Male External Genitals

Skin of Scrotum and Penis

The cutaneous lymphatics of the scrotum and the penis anastomose with each other and with the perineal lymphatics. The lymph vessels of the internal and external surface of the prepuce converge toward the dorsal aspect, join with the lymphatics of the skin of the shaft and form several collecting lymph vessels which ascend to the pubic region, follow the femoral canal and terminate in the superficial inguinal lymph nodes. Rarely, anastomoses with the lower inguinal nodes or the contralateral inguinal nodes can be demonstrated.

Regional lymph nodes   *Lnn. inguinales superficiales*

Glans and Body of Penis

The rich lymphatics of the glans converge toward the frenulum, where they anastomose with the lymphatics of the urethra and give rise to several collecting lymph vessels. These vessels follow the retroglandular sulcus and then run along the dorsal surface of the penis under the penile fascia and in close proximity to the deep dorsal vein. At the suspensory ligament they form a presymphyseal plexus with occasional lymph nodes. Then, they usually pass through the femoral canal and empty into the homolateral superficial inguinal nodes. Less frequently, they pass through the inguinal canal and terminate in the lateral lacunar node, through the femoral canal and end in the medial lacunar node or drain to the deep inguinal lymph nodes [21].

The same lymphatic drainage may be demonstrated for the corpora cavernosa.

Regional lymph nodes   *Lnn. inguinales superficiales et profundi*
                            Lnn. iliaci externi
                            (Ln. lacunaris medialis et lateralis)

Female Internal Genitals

Ovary (Fig. 56)

The ovary possesses an extremely rich lymphatic network which surrounds the graafian follicles. This network gives rise to six to eight collecting vessels which form the "subovarian plexus" [44] in the mesovarium. The main lymph vessels leaving the plexus

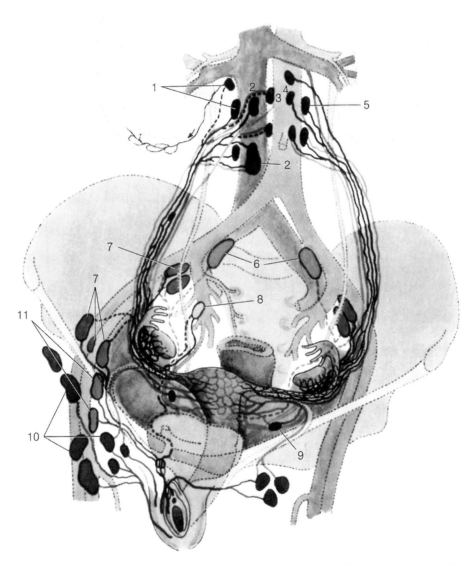

Fig. 56. The lymphatic drainage of the ovaries, the fallopian tubes, the body of the uterus and the external genitals. (Adapted from [23])

1 Laterocaval lymph nodes
2 Precaval lymph nodes
3 Intermediate lumbar lymph nodes
4 Pre-aortic lymph nodes
5 Latero-aortic lymph nodes
6 Common iliac lymph nodes
7 External iliac lymph nodes
8 Internal iliac lymph nodes
9 Parauterine lymph nodes
10 Superficial inguinal lymph nodes
11 Deep inguinal lymph nodes

accompany the utero-ovarian vessels, cross the ureter and the external iliac vessels and, lying on the psoas muscle, ascend to the level of the lower pole of the kidney [36]. There, they cross medially in front of the ureter, diverge like a fan and terminate in lumbar nodes between the right renal and the inferior mesenteric artery. This applies to the right side; on the left, the lymph vessels do not spread out in this way. Instead the main vessels terminate in pre- and latero-aortic nodes and in lumbar nodes located in the vicinity of the left renal artery and vein [21]. Occasionally, efferent vessels on the right terminate in subaortic nodes.

Inconstantly, a collecting lymph vessel may run laterally in the suspensory ligament of the ovary to terminate in the intermediate external iliac nodes.

Regional lymph nodes   *Lnn. lumbales dextri et sinistri*
                                     Lnn. iliaci externi intermedii

Body of Uterus, Fallopian Tube (Fig. 56)

Both organs dispose of a lymphatic network in the mucosa, the muscularis and the subserosa, the latter giving rise to collecting lymph vessels on the lateral border of the uterus which run in the suspensory ligament along the ovarian branch of the uterine artery. Further on, these vessels join the tubo-ovarian lymph vessels and ascend to the lumbar nodes.

Less frequently, some collecting lymph vessels run in a transverse direction to external iliac lymph nodes. Their course may be interrupted by parauterine nodes. Occasionally, a separate lymph vessel from the ampulla of the fallopian tube runs along the broad ligament to reach the superior gluteal nodes. Some lymph vessels may be found to originate at the insertion of the round ligament of the uterus and then follow this ligament and empty into the superomedial group of superficial inguinal nodes [21].

Apart from the communications with the ovarian lymphatics, the lymphatics of the fallopian tube and the body of the uterus anastomose not only peripherally, but also intramurally.

Regional lymph nodes   *Lnn. lumbales dextri et sinistri*
                                     Lnn. iliaci externi
                                     Lnn. iliaci interni
                                     (Lnn. gluteales superiores)
                                     Lnn. inguinales superficiales
                                     (Lnn. inguinales superficiales superomediales)
                                     Lnn. para-uterini

Uterine Cervix

An abundant lymphatic network may be demonstrated in the parametrium of the uterine cervix. This lymphatic plexus gives rise to three groups of main collecting lymph vessels [21, 36]:

1. The group of anterior or preureteral collecting vessels represents the principal lymphatic chain of the cervix. It passes the broad ligament of the uterus, runs in close proximity to the uterine artery, passes in front of the ureters, crosses the umbilical artery and empties to the upper and middle nodes of the intermediate and medial external iliac chains. Inconstantly, lymph vessels may drain to the obturator group of the external iliac nodes.
2. The group of retroureteral vessels accompanies the uterine vein, passes behind the ureter and terminates in the internal iliac nodes and sometimes in the promontory nodes.
3. The posterior or uterosacral group of collecting lymph vessels runs posteriorly on each side of the rectum and then sweeps upward to the lateral sacral, the promontory and the subaortic nodes.

The lymph of the uterine cervix is collected by these main collecting lymph vessels either directly or after passing through the parauterine and paravaginal lymph nodes. The largest of those lymph nodes is located at the intersection of the ureter and the uterine artery and is called the ureterouterine node.

The lymphatics of the cervix anastomose intramurally and in the parauterine space with the lymph vessels of the body of the uterus and the vagina.

Regional lymph nodes   *Lnn. iliaci externi*
(Lnn. iliaci externi mediales et intermedii)
Lnn. iliaci interni
(Lnn. sacrales laterales)
Lnn. iliaci communes
(Lnn. iliaci communes subaorttci et promontorii)
Lnn. para-uterini
Lnn. paravaginales

Vagina

The vagina disposes of a rich lymphatic network in the mucosa, the submucosa and the muscular layers which gives rise to collecting lymph vessels laterally to the organ originating on the anterior and posterior vaginal wall [36].

The collecting lymph vessels of the upper half of the vagina accompany the vaginal and uterine arteries, inconstantly may pass the ureterouterine node, and drain to the middle and upper nodes of the medial and intermediate nodes of the external iliac chain. Sometimes, the lymph is directed to the medial lacunar, the obturator, the superior gluteal and the pararectal lymph nodes.

The lower half of the vagina is drained indirectly by paravaginal and parauterine lymph nodes or directly by collecting lymph vessels which lie in close relationship to the vaginal artery, cross the blood vessels of the upper vagina and empty to the internal iliac nodes. Additionally, there are inconstant communications with the superior and inferior gluteal and promontory lymph nodes.

The separation of these two main drainage regions of the vagina is somewhat artificial, as communications between the regions exist. In addition, anastomoses to the lymphatics of the vulva, the cervix and the rectum have been demonstrated [21].

Regional lymph nodes   *Lnn. iliaci externi*
(Lnn. iliaci externi mediales, intermedii et obturatorii,
Ln. lacunaris medialis)
Lnn. iliaci interni
(Lnn. gluteales superiores et inferiores)
Lnn. iliaci communes
(Lnn. iliaci communes promontorii)
Lnn. paravaginales
Lnn. para-uterini
Lnn. pararectales

Female External Genitals

Vestibule of Vagina, Labia Pudendi (Fig. 56)

The fine lymphatic network of the labia pudendi (except the lateral surface of the labia majora) and the prepuce of the clitoris gives rise to collecting lymph vessels which predominantly drain to the homolateral superficial inguinal lymph nodes.

Regional lymph nodes   *Lnn. inguinales superficiales*

Body and Glans of Clitoris, Bartholin's Glands (Fig. 56)

The collecting lymph vessels of the body and the glans of the clitoris run to the presymphyseal plexus, which may contain a small lymph node. From there, the main lymph vessels drain either to the deep inguinal lymph nodes and the medial lacunar node or to the lateral lacunar node via the inguinal canal and along the round ligament of the uterus. Rarely, lymph vessels may accompany the internal pudendal or the vesical artery and reach the internal iliac nodes.

As for Bartholin's glands, few data on lymphatic drainage are available. There seem to be lymphatic pathways to the superficial inguinal nodes as well as to the inferior gluteal and other internal iliac nodes [36].

Regional lymph nodes   *Lnn. inguinales superficiales et profundi*
                              Lnn. iliaci externi
                              (Ln. lacunaris mediales et laterales)
                              Lnn. iliaci interni

Topographical Anatomy of Regional Lymph Node Groups of Pelvis and Inguinal Region (Figs. 57-66)

Parietal Nodes

Lnn. Rectales Superiores

According to the most recent edition of the *Nomina Anatomica* [32], these nodes belong to the inferior mesenteric chain of the abdominal visceral nodes, but, in fact, they represent regional lymph nodes for the rectum. They are grouped around the superior rectal artery in the mesorectum. The so-called principal node of the rectum may be found at the bifurcation of the superior rectal artery.

Lnn. Iliaci Communes (Figs. 57, 58)

These nodes are situated along the common iliac vessels and represent the secondary lymph draining nodes for the pelvic and genital organs, the inner surface of the pelvic wall, the lower abdominal wall up to the umbilicus and the gluteal and iliac muscles. The efferent lymph flow finds its way to the lumbar nodes and the lumbar trunk. The common iliac nodes are subdivided into several groups.

Lnn. iliaci communes mediales: medial to the common iliac vessels.

Lnn. iliaci communes intermedii: behind the vessels.

Lnn. iliaci communes laterales: lateral to the vessels.

Lnn. iliaci communes subaortici: just below the aortic bifurcation in front of the fourth lumbar vertebra.

Lnn. iliaci communes promontorii (Fig. 58): on the inner surface of the sacral promontory.

Lnn. Iliaci Externi (Figs. 59–61, 64, 65)

These lymph nodes are grouped around the external iliac vessels and represent the primary lymph drainage nodes for parts of the bladder and the vagina and the secondary nodes for the lymph of the inguinal nodes. The efferent lymph vessels drain to the common iliac chain. The external iliac nodes consist of several groups.

Lnn. iliaci externi mediales: medial to the external iliac vessels.

Lnn. iliaci externi intermedii: between the medial and the lateral group just behind the artery.

Lnn. iliaci externi laterales: lateral to the vessels.

Lnn. iliaci externi interiliaci: at the branching of the common iliac artery into the internal and the external artery.

Lnn. iliaci externi obturatorii: in close relationship to the obturator artery.

Ln. lacunaris medialis: a single node medial to the vessels in the lacuna vasorum.

Ln. lacunaris intermedius: an inconstant node in the middle of the lacuna vasorum.

Ln. lacunaris lateralis: a node lateral to the vessels in the lacuna vasorum.

Lnn. Iliaci Interni (Figs. 59, 64)

These lymph nodes are situated around the internal iliac artery and drain the lymph from the pelvic organs, the inner and outer pelvic wall and the perineum. They can be subdivided into three groups.

Lnn. gluteales superiores: nodes for the pelvic wall, in close proximity to the superior gluteal artery.

Lnn. gluteales inferiores: nodes for the prostate and the proximal urethra, along the inferior gluteal artery.

Lnn. sacrales (Figs. 59, 60): in front of the sacrum, draining the lymph of the prostate and the cervix uteri.

Visceral Nodes

Lnn. Paravesiculares

These lymph nodes lie in close relationship to the bladder and receive lymph from the bladder and the prostate. There are three subgroups.

Lnn. prevesiculares (Fig. 61): in the space between bladder and symphysis.

Lnn. postvesiculares (Figs. 60, 65): just behind the dorsal wall of the bladder.

Lnn. vesiculares laterales (Figs. 60, 65): at the lower end of the medial umbilical ligament.

Lnn. Para-uterini (Figs. 64, 65)

This group is situated lateral to the uterus and drains especially the lymph of the cervix.

Lnn. Paravaginales (Fig. 66)

These lie lateral to the vagina and receive lymph from some parts of it.

Lnn. Pararectales (Anorectales) (Figs. 59–62, 64–66)

This group drains lymph from the rectum and some parts of the vagina and can be found on the lateral wall of the rectum.

Inguinal Nodes

Lnn. Inguinales Superficiales (Figs. 62, 63, 66)

This group is found in the subcutaneous tissue on the fascia lata and drains lymph of the anal canal, the perineum, the external genitals, the abdominal wall and the leg. The efferent lymph vessels terminate in the external iliac chain. The superficial inguinal nodes consist of three subgroups.

Lnn. inguinales superficiales superomediales: in the medial part of the inguinal ligament.

Lnn. inguinales superficiales superolaterales: lateral to the medial group along the inguinal ligament.

Lnn. inguinales superficiales inferiores: grouped around the terminal part of the long saphenous vein.

Lnn. Inguinales Profundi (Figs. 62, 63, 66)

This group lies at the saphenous opening of the fascia lata and drains to the external iliac nodes.

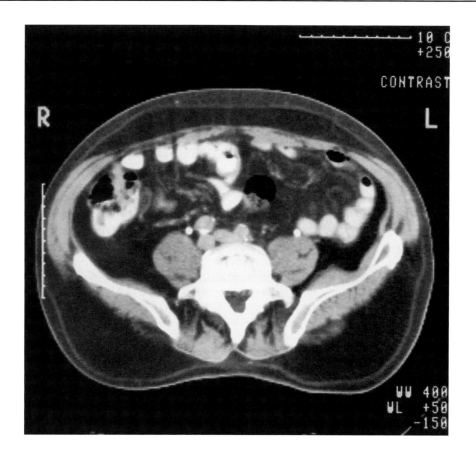

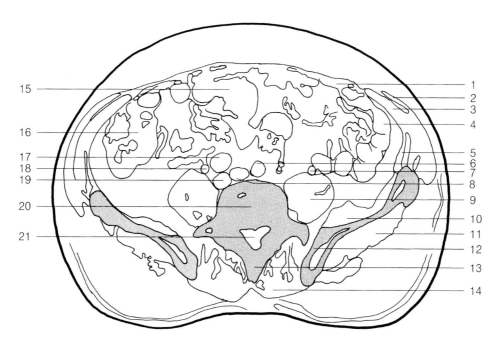

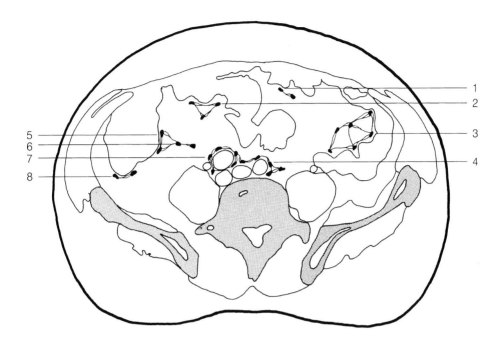

Fig. 57. Axial section of the pelvis
(male subject)   ▲

◄ 1 Abdominis muscle
   2 External oblique muscle
   3 Internal oblique muscle
   4 Colon
   5 Transverse muscle of abdomen
   6 Left common iliac artery
   7 Ureter
   8 Left common iliac vein
   9 Greater psoas muscle
  10 Iliac muscle
  11 Gluteus medius muscle
  12 Ala of ilium
  13 Spine of a vertebra
  14 Erector muscle of spine
  15 Ileum
  16 Colon
  17 Right common iliac artery
  18 Ureter
  19 Right common iliac vein
  20 Vertebra
  21 Vertebral canal

1–3 Juxtaintestinal lymph nodes
  4 Common iliac lymph nodes
     (medial, intermediate, lateral)
  5 Paracolic lymph nodes
  6 Colonic lymph nodes
  7 Common iliac lymph nodes
  8 Retrocecal lymph nodes

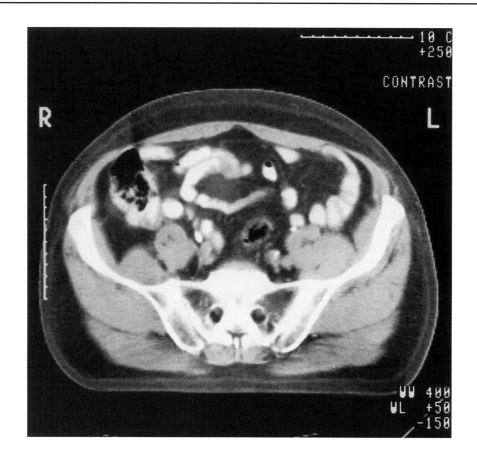

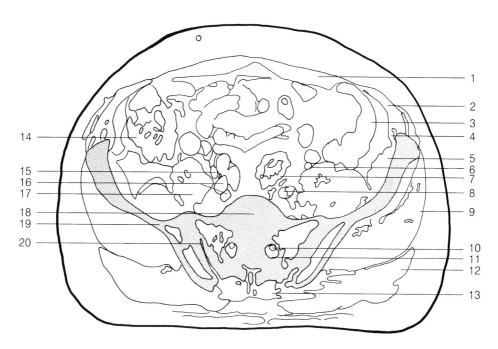

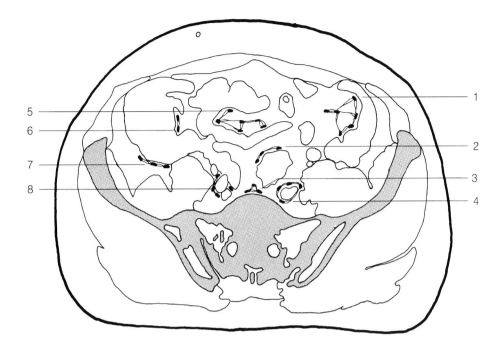

Fig. 58. Axial section of the pelvis
(male subject)

▲

◄ 1 Rectus abdominis muscle
    2 External + internal oblique muscles
    3 Ileum
    4 Transverse muscle of abdomen
    5 Iliac muscle
    6 Ureter
    7 Sigmoid colon
    8 Common iliac artery
    9 Gluteus medius muscle
   10 Root of a spinal nerve
   11 Sacral foramen
   12 Gluteus maximus muscle
   13 Erector muscle of spine
   14 Ascending colon
   15 Ureter
   16 Common iliac artery and vein
   17 Greater psoas muscle
   18 Promontory of sacrum
   19 Ala of ilium
   20 Sacroiliac joint

1 Juxtaintestinal lymph nodes
2 Sigmoid lymph nodes
3 Common iliac promontory lymph nodes
4 Common iliac lymph nodes
   (medial, intermediate, lateral)
5 Juxtaintestinal lymph nodes
6 Paracolic lymph nodes
7 Retrocaval lymph nodes
8 Common iliac lymph nodes

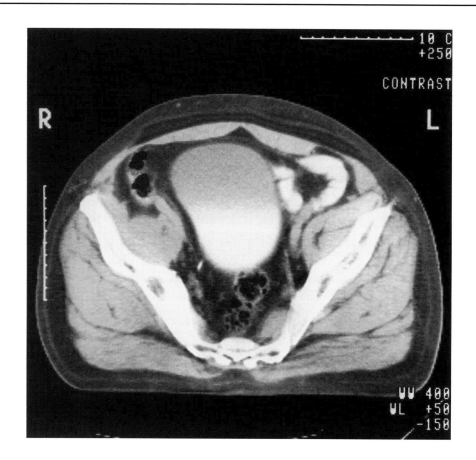

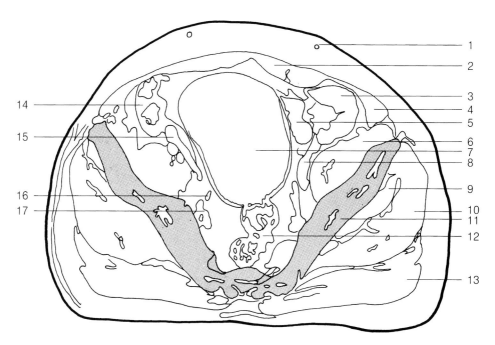

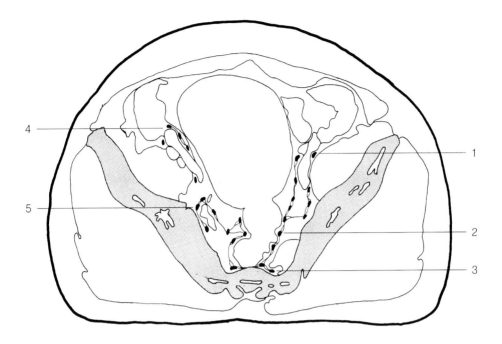

Fig. 59. Axial section of the pelvis
(male subject)

▲

◀ 1 Inferior epigastric artery
   2 Rectus abdominis muscle
   3 Sigmoid colon
   4 Internal oblique muscle
   5 Ileum
   6 Iliopsoas muscle
   7 Urinary bladder
   8 External iliac artery and vein
   9 Gluteus minimus muscle
  10 Gluteus medius muscle
  11 Ala of ilium
  12 Rectum
  13 Gluteus maximus muscle
  14 Ascending colon
  15 External iliac artery and vein
  16 Ureter
  17 Internal iliac artery and vein

   1 External iliac nodes
   2 Pararectal lymph nodes
   3 Sacral lymph nodes
   4 External iliac lymph nodes
   5 Internal iliac lymph nodes

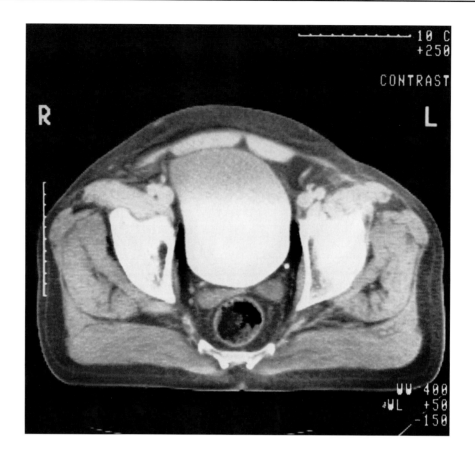

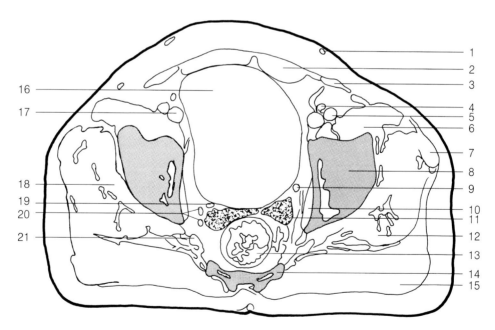

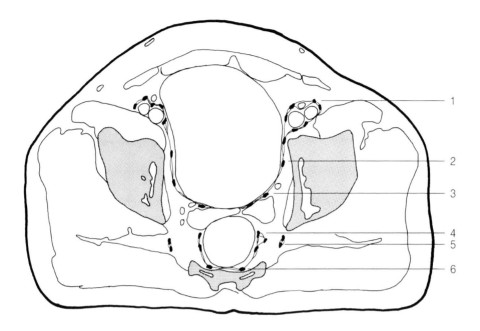

Fig. 60. Axial section of the pelvis
(male subject)   ▲

◄ 1 Inferior epigastric artery
   2 Rectus abdominis muscle
   3 Internal oblique muscle
   4 Deep circumflex iliac artery and vein
   5 External iliac artery
   6 Iliopsoas muscle
   7 Tensor muscle of fascia lata
   8 Acetabulum
   9 Ureter
  10 Internal obturator muscle
  11 Seminal vesicle
  12 Gluteus medius muscle
  13 Rectum
  14 Sacrum
  15 Gluteus maximus muscle
  16 Urinary bladder
  17 External iliac vein
  18 Gluteus minimus muscle
  19 Ureter
  20 Inferior vesical artery and vein
  21 Internal pudendal artery and vein

1 External iliac lymph nodes
2 Laterovesical lymph nodes
3 Retrovesical lymph nodes
4 Pararectal lymph nodes
5 Gluteal lymph nodes
6 Sacral lymph nodes

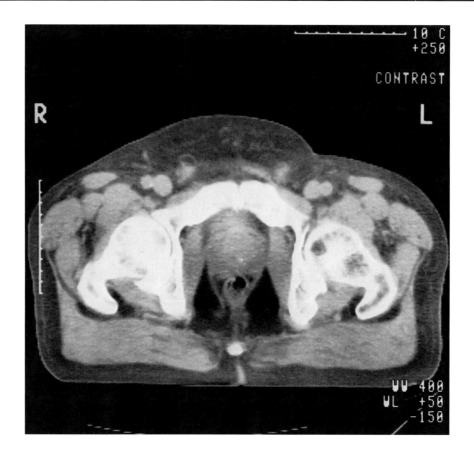

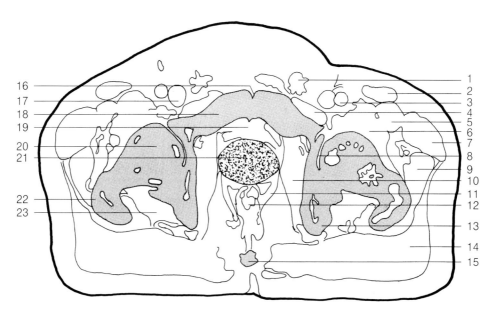

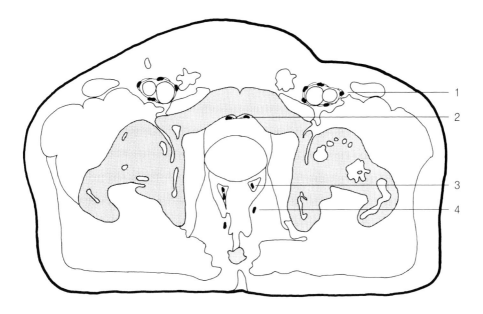

Fig. 61. Axial section of the pelvis
(male subject)

◄ 1 Spermatic cord
   2 Sartorius muscle
   3 External iliac artery
   4 Pectineal muscle
   5 Rectus femoris muscle
   6 Iliopsoas muscle
   7 Tensor muscle of fascia lata
   8 Hip joint
   9 Gluteus medius muscle
  10 Internal obturator muscle
  11 Levator ani muscle
  12 Rectum
  13 Body of ischium
  14 Gluteus maximus muscle
  15 Coccyx
  16 Pubic symphysis
  17 External iliac vein
  18 Superior pubic ramus
  19 Urinary bladder
  20 Head of femur
  21 Prostate
  22 Greater trochanter
  23 Gemellus inferior muscle

▲

  1 External iliac lymph nodes
  2 Prevesical lymph nodes
3, 4 Pararectal lymph nodes

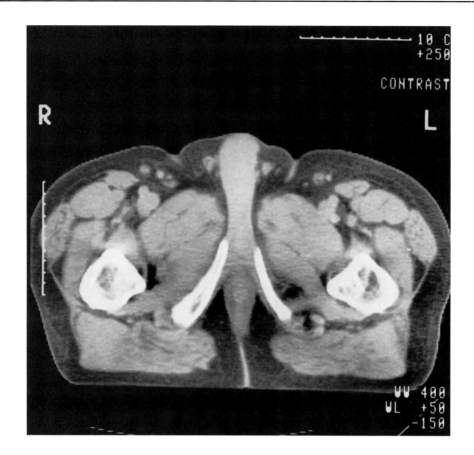

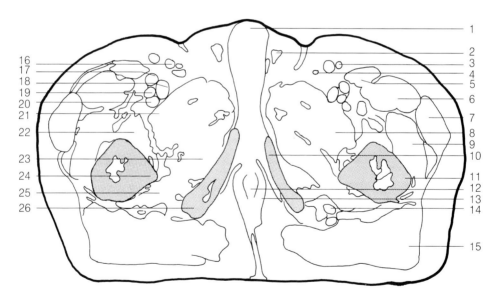

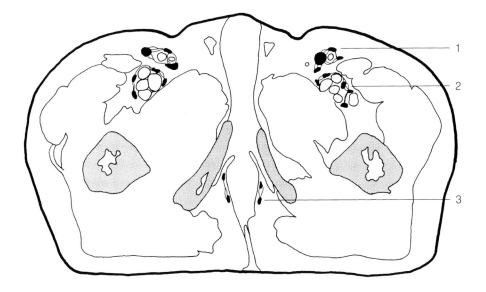

Fig. 62. Axial section of the pelvis (male subject)

◀

 1 Corpus cavernosum
 2 Spermatic cord
 3 Long saphenous vein
 4 External pudendal vein
 5 Sartorius muscle
 6 Rectus femoris muscle
 7 Tensor muscle of fascia lata
 8 Short adductor muscle
 9 Vastus lateralis muscle
10 Ramus of ischium
11 Femur
12 Rectum
13 Internal sphincter muscle of anus
14 Sciatic nerve
15 Gluteus maximus muscle
16 Long saphenous vein
17 Femoral artery
18 Femoral vein
19 Deep femoral artery
20 Deep femoral vein
21 Pectineal muscle
22 Iliopsoas muscle
23 Great adductor muscle
24 Lesser trochanter
25 Quadratus femoris muscle
26 Ischial tuberosity

▲

1 Superficial inguinal lymph nodes
2 Deep inguinal lymph nodes
3 Anorectal lymph nodes

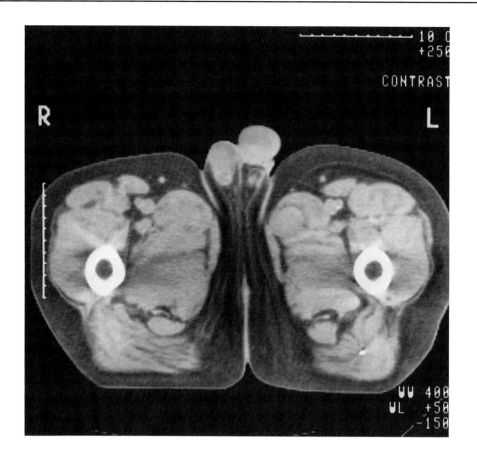

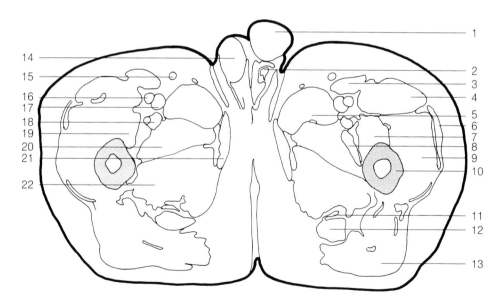

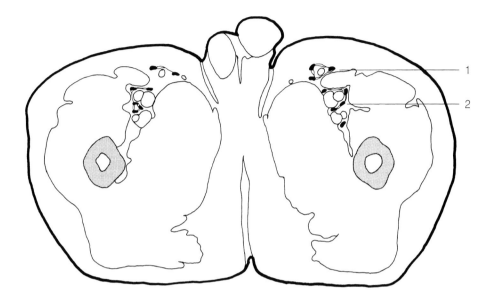

Fig. 63. Axial section of the pelvis
(male subject)

▲

◄ 1 Testicle
   2 Ductus deferens
   3 Sartorius muscle
   4 Rectus femoris muscle
   5 Long adductor muscle
   6 Tensor muscle of fascia lata
   7 Vastus intermedius muscle
   8 Pectineal muscle
   9 Vastus lateralis muscle
  10 Femur
  11 Semimembranous muscle
  12 Semitendinous muscle
  13 Gluteus maximus muscle
  14 Testicle
  15 Long saphenous vein
  16 Femoral artery
  17 Femoral vein
  18 Deep femoral artery and vein
  19 Vastus medialis muscle
  20 Short adductor muscle
  21 Gracilis muscle
  22 Great adductor muscle

1 Superficial inguinal lymph nodes
2 Deep inguinal lymph nodes

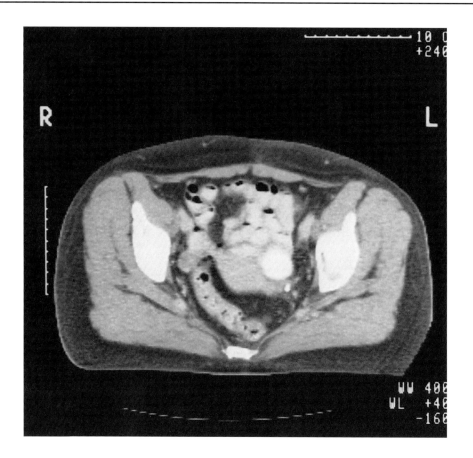

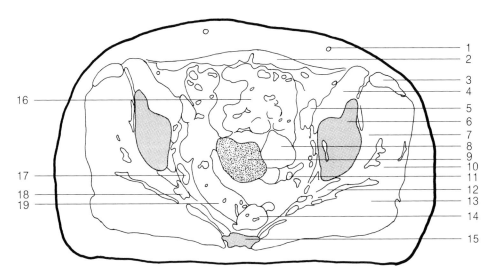

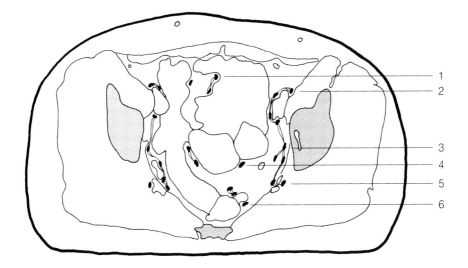

Fig. 64. Axial section of the pelvis
(female subject)

▲

◄ 1 Superficial epigastric vein
    2 Rectus abdominis muscle
    3 Tensor muscle of fascia lata
    3 Iliopsoas muscle
    4 External iliac artery and vein
    5 Hip bone
    6 Gluteus minimus muscle
    8 Urinary bladder
    9 Uterus
  10 Gluteus medius muscle
  11 Ureter
  12 Internal obturator muscle
  13 Gluteus maximus muscle
  14 Rectum
  15 Coccyx
  16 Ileum
  17 Internal iliac artery and vein
  18 Inferior gluteal artery and vein
  19 Ileum

1 Juxtaintestinal lymph nodes
2 External iliac lymph nodes
3 Internal iliac lymph nodes
4 Parauterine lymph nodes
5 Gluteal lymph nodes
6 Pararectal lymph nodes

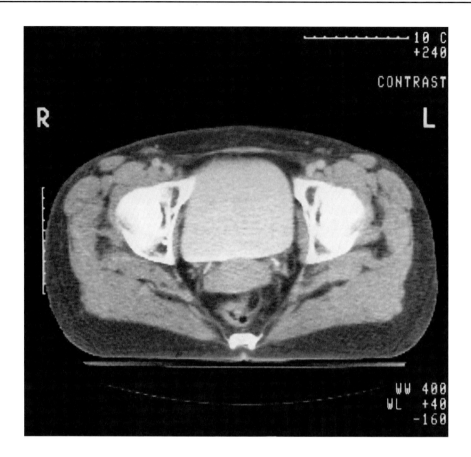

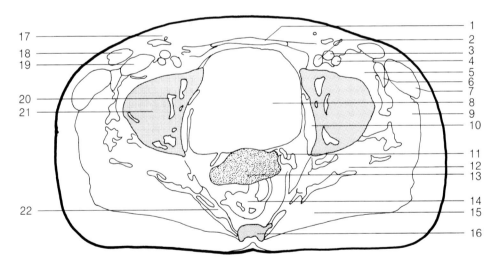

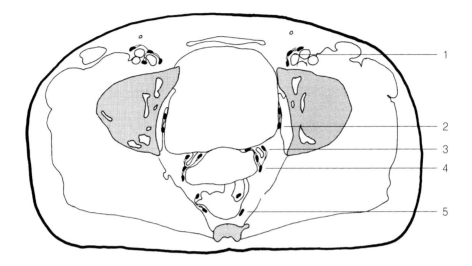

Fig. 65. Axial section of the pelvis
(female subject)

▲

◄ 1 Pyramidal muscle
   2 External iliac vein
   3 External iliac artery
   4 Femoral nerve
   5 Iliopsoas muscle
   6 Rectus femoris muscle
   7 Tensor muscle of fascia
   8 Urinary bladder
   9 Gluteus medius muscle
  10 Internal obturator muscle
  11 Ureter
  12 Gluteus minimus muscle
  13 Uterus
  14 Rectum
  15 Gluteus maximus muscle
  16 Coccyx
  17 Superficial epigastric vein
  18 Sartorius muscle
  19 Iliopsoas muscle
  20 Hip Joint
  21 Head of femur
  22 Sacrotuberous ligament

1 External iliac lymph nodes
2 Laterovesical lymph nodes
3 Retrovesical lymph nodes
4 Parauterine lymph nodes
5 Pararectal lymph nodes

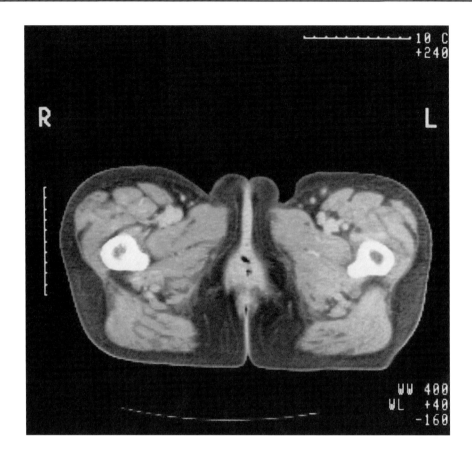

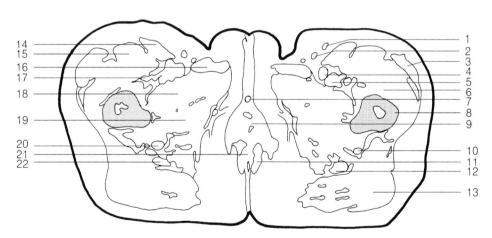

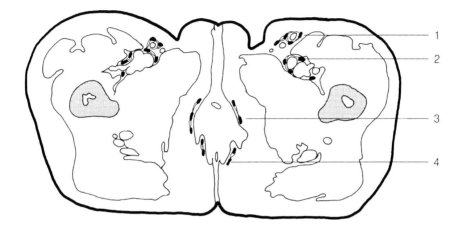

Fig. 66. Axial section of the pelvis
(female subject)

▲

◄ 1 Long saphenous vein
  2 External pudendal vein
  3 Tensor muscle of fascia lata
  4 Femoral artery
  5 Femoral vein
  6 Vastus intermedius muscle
  7 Urethra
  8 Shaft of femur
  9 Vagina
10 Sciatic nerve
11 Internal sphincter muscle of anus
12 Semitendinous muscle
13 Gluteus maximus muscle
14 Sartorius muscle
15 Rectus femoris muscle
16 Long adductor muscle
17 Lateral circumflex femoral artery and vein
18 Short adductor muscle
19 Long adductor muscle
20 Tendon of semimembranous muscle
21 Anus
22 Semitendinous muscle

1 Superficial inguinal lymph nodes
2 Deep inguinal lymph nodes
3 Paravaginal lymph nodes
4 Anorectal lymph nodes

References

1. Ackermann LV, del Regato JA (1970) Cancer – diagnosis, treatment, and prognosis, 4th edn. Mosby, St. Louis, pp 378, 554, 575
2. Arao A, Abrao A (1954) Esdudo anatomico da cadeia ganglionar mamaria interna em 100 casos. Rev paulista de med 45: 317
3. Asellius G (1627) De lacteibus sive lactus remis quarto vasorum mesaroicum geneve novo invente. Mediolami, Bidellus
4. Bartels PC (1909) Das Lymphgefäßsystem. In: Bardleben K (ed) Handbuch der Anatomie, vol 3. Fischer, Jena
5. Caplan I (1982) Drainage lymphatique intra- et extra-hepatique de la vesicule biliaire. Bull Mem Acad R Med Belg 137: 324–334
6. Coller FA, Kay EB, McIntyre RS (1941) Regional lymphatic metastases of carcinoma of the stomach. Arch Surg 43: 748–761
7. Daehm H (1959) Contribution à l'étude des voies lymphatiques du larynx et de la trachée. Acta Otorhinolaryng Belg 13: 229
8. Deki H, Sato T (1988) An anatomic study of the peripancreatic lymphatics. Surg Radiol Anat 10: 121–135
9. Delamere G, Poirier P, Cunéo B (1913) The lymphatics. Constable, London
9a. del Regato JA, Spjut HA, Cox JD (1985) Ackerman and del Regato's cancer: diagnosis, treatment, and prognosis, 6th edn. Mosby, St. Louis
10. De Sousa OM (1954) Sur les variations du drainage lymphatique du lobe inférieur du poumon chez l'homme. Acta Anat 21: 342–348
11. Feneis H (1988) Pocket atlas of human anatomy. Thieme, Stuttgart
12. Fisch UG, Sigl ME (1964) Cervical lymphatic system as visualized by lymphography. Ann Otol Rhinol Laryngol 73: 869–882
13. Fletcher GH (1980) Textbook of radiotherapy, Lea & Febiger, Philadelphia, pp 286, 669, 897
14. Gall FP, Scheele J (1986) Maligne Tumoren des Rektums. In: Gall FP, Hermanek P, Tonak J (eds) Chirurgische Onkologie. Springer, Berlin Heidelberg New York, p 524
15. Haagensen CD (1972) Methods of study of the lymphatics system. In: Haagensen CD et al. (eds) The lymphatics in cancer. Saunders, Philadelphia
16. Hollinshead WH (1968) Anatomy for surgeons, vol 2. Hoeber Medical, Harper & Row, New York, p 15
17. Husemann B (1986) Maligne Tumoren des Oesophagus. In: Gall FP, Hermanek P, Tonak J (eds) Chirurgische Onkologie. Springer, Berlin Heidelberg New York, p 327
18. Kett K, Varga G, Lukacs L (1970) Direct lymphography of the breast. Lymphology 3: 3
19. Kinmonth JBV (1952) Lymphography in man: method of outlining lymphatic trunks at operation. Clin Sci 11: 13
20. Kubik S (1974) The anatomy of the lymphatic system. In: Musshoff K (ed) Recent results in cancer research, vol 46. Springer, New York Heidelberg Berlin, pp 5–17
21. Kubik S (1980) Visceral lymphatic system. In: Viamonte M, Rüttimann A (eds) Atlas of lymphography. Thieme, Stuttgart, pp 91–106
22. Kubik S (1981) Anatomie des Lymphgefäßsystems. In: Frommhold W, Gerhardt P (eds) Klinisch-radiologisches Seminar. Erkrankungen des Lymphsystems. Thieme, Stuttgart
23. Kubik S, Wirth W (1980) Histology, anatomy and lymphographic appearance. In: Viamonte M, Rüttimann A (eds) Atlas of lymphography, Thieme, Stuttgart 1–17

24. Kubik S, Töndery G, Rüttimann A, Wirth W (1967) Nomenclature of the lymph nodes of the retroperitoneum, the pelvis and the lower extremity. In: Rüttimann A (ed) Progress in lymphology. Thieme, Stuttgart, plates I, II, IV.

25. Mann W (1979) Das Lymphsystem des Kehlkopfes, eine lymphangioskopische und ultrastrukturelle Untersuchung. Arch Otorhinolaryngol 225: 165

26. Mascagni P (1787) Vasorum lymphaticorum corporis humani historia et iconographia. Senis (Siena) ex typographia Pazzini Carli

27. Meister R (1986) Maligne Lungentumoren. In: Gall FP, Hermanek P, Tonak J (eds) Chirurgische Onkologie. Springer, Berlin Heidelberg New York, p 250

28. Most A (1899) Über die Lymphgefäße und Lymphdrüsen des Kehlkopfes. Anat Anz 15: 387

29. Most A (1906) Die Topographie des Lymphgefäßapparates des Kopfes und des Halses in ihrer Bedeutung für die Chirurgie. Hirschwald, Berlin

30. Naumann HH (1972) Chirurgie der malignen Tumoren des Larynx. In: Naumann HH (ed) Kopf- und Halschirurgie, vol 1. Thieme, Stuttgart, p 189

31. Nohl-Oser HC (1972) Lymphatics of the lung. In: Shields TW (ed) General thoracic surgery. Lea & Febiger, Philadelphia, pp 72–81

32. Nomina anatomica, 5th edn (1983) Williams & Wilkins, Baltimore

33. Pecquet J (1653) New anatomical experiments. O. Pulleyn London

34. Pissas A (1984) Anatomicoclinical and anatomicosurgical essay on the lymphatic circulation of the pancreas. Anat Clin 6: 255–280

35. Pissas A, Dyon JF, Sarrazin R, Boucher Y (1979) Le drainage lymphatique de l'estomac. J Chir 116: 583–590

36. Plentl AA, Friedman EA (1971) Lymphatic system of the female genitalia. Saunders, Philadelphia

37. Poirier P, Cunéo B (1909) Les lymphatiques, tome deuxieme. In: Poirier, Charpy (eds) Traité de l'anatomie humaine. Masson, Paris

38. Rawson AJ (1949) Distribution of the lymphatics of the human kidney as shown in a case of carcinomatous permeation. Arch Path 47: 283–288

39. Resano JH (1951) Traitement chirurgical du cancer de segment juxta-hilaire de l'oesophage. Presse med 59: 1200–1204

40. Rotter J (1899) Zur Topographie des Mammacarcinoms. Arch Klin Chir 58: 346–356

41. Roubaud L (1902) Contribution à l'étude anatomique des lymphatiques du larynx. Thesis, University of Paris

42. Rouvière H (1932) Anatomie des lymphatiques de l'homme. Masson, Paris

43. Sabin FR (1901) On the origin of the lymphatic system from the veins and the development of the lymph hearts and thoracic duct in the pig. Am J Anat 1: 367

44. Sappey PC (1885) Description et iconographic des vaisseaux lymphatiques considerés chez l'homme et les vertébrés. A. Délahaye and E. Lecroisnier, Paris

45. Shdanov DA (1932) Röntgenologische Untersuchungsmethoden des Lymphgefäßsystems des Menschen und der Tiere. Fortschr. Röntgenstr. 46: 680

46. Sukiennikow W (1903) Topographische Anatomie der bronchialen und trachealen Lymphdrüsen. Berl Klin Wochenschr 40: 369–372

47. Taillens JP (1960) Etude anatomo-clinique des chaines ganglionnaires lymphatiques du cou et de leurs ganglions satellites bucco-pharyngés. Pract Otorhinolaryng 22: 44

Bibliography Update

Aquino SL, Hayman LA, Loomis SL, Taber KH (2003) Source and direction of thoracic lymphatics, part I: The upper thorax. J Comput Assist Tomogr 27:292-6

Aquino SL, Hayman LA, Loomis SL, Taber KH (2003) The source and direction of thoracic lymphatics, part II: The lower thorax. J Comput Assist Tomogr 27:657-61

Brossner C, Ringhofer H, Schatzl G, Madersbacher S, Powischer G, Kuber W (2002) Sacral distribution of prostatic lymph nodes visualized on spinal computed tomography with three-dimensional reconstruction. BJU Int 89:44-7

Dorfman RE, Alpern MB, Gross BH, Sandler MA (1991) Upper abdominal lymph nodes: criteria for nomal size determined with CT. Radiology 180:319-22

Fishman EK, Zinreich ES, Jacobs CG, Rostock RA, Siegelman SS (1986) CT of the axilla: normal anatomy and pathology. Radiographics 6:475-502

Földi M, Kubik S (1999) Lehrbuch der Lymphologie 4. Auflage

Genereux GP, Howie JL (1984) Normal mediastinal lymph node size and number: CT and anatomic study. AJR Am J Roentgenol 142:1095-100

Iyer RB, Libshitz HI (1995) Radiographic demonstration of intercostal lymphatics and lymph nodes. Lymphology 28:89-94

Konig R, van Kaick G (1982) Computer tomographic differentiation of lymph nodes in the retroperitoneum and lower mediastinum, with special reference to normal anatomic structures. Computertomographie 2:184-9

Lien HH, Lund G (1985) Computed tomography of mediastinal lymph nodes. Anatomic review based on contrast enhanced nodes following foot lymphography. Acta Radiol Diagn (Stockh.) 26:641-7

Murray JG, Breatnach E (1993) The American Thoracic Society lymph node map: a CT demonstration. Eur J Radiol 17:61-8

Peters PE, Beyer K (1985) Normal lymph node cross sections in different anatomic region and their significance for computed tomographic diagnosis. Radiologe 25:193-8

Roberts KT, Mettler FA Jr (1979) Diagnostic evaluation of the pelvic and abdominal lymphatic system. Curr Probl Diagn Radiol 8:1-56

Subject Index

bronchomediastinal trunk
  s. lymphatic trunk

cisterna chyli *Fig. 38*, 1, 3, 6,
  41, 44, 45, 48, 82, 88, 93
computed tomography (CT) 3

intestinal trunk s. lymphatic
  trunk

jugular trunk s. lymphatic
  trunk

lumbar trunk s. lymphatic
  trunk
lymphatic drainage
– abdominal wall *82*, 127
– anal canal *Fig. 52*, 114,
  115, 12
– Bartholin's glands *Fig. 56*,
  124, 125
– basic principle *Fig. 2*, 3
– bile ducts, extrahepatic *84*
– breast *Fig. 23*, 48, 49
– buccal mucosa s. face
– chin s. face
– clitoris, body and glans
  Fig. 56, 124, 125
– colon *Fig. 42*, 88, 89
– conjunctiva s. orbit
– costal wall *48*
– diaphragm *48*
– ductus deferens 117, 118,
  119
– duodenum s. small bowel
– ear *Fig. 5*, 11
– epididymis *Fig. 55*, 118,
  119
– esophagus *Fig. 22*, 46, 47
– eustachian tube s. ear and
  nasopharynx
– eyelid s. face
– face *Fig. 5*, 9, 129
– fallopian tube *Fig. 56*,
  114, 122

– floor of the mouth s. oral
  cavity gallbladder 84
– gingiva of the mandible
  s. oral cavity
– heart 41, 46
– hypopharynx 13
– kidney 90
– labia pudendi s. vulva
– lacrimal gland s. face
– larynx *Fig. 7*, 7, 13, 14
– lip s. face
– liver *Fig. 39*, 82, 83
– lung *Figs. 19, 20*, 41–43
– maxillary sinus s. paranasal
  sinuses
– mediastinum 44, 45
– nasopharynx 7, 13
– nose 10
– oral cavity 12
– orbit 11
– oropharynx 7, 13
– ovary *Fig. 56*, 114, 121,
  122
– palate, hard and soft s. oral
  cavity
– pancreas *Fig. 41*, 86, 87
– paranasal sinuses 7, 11
– parotid gland *Fig. 5*, 15
– penis, body and glans 120
– pericardium 46
– perineum 114, 126, 127
– pleura, parietal 48
– prostate *Figs. 53, 54*, 114,
  117, 118, 126
– rectum *Fig. 52*, 114, 115,
  124, 127
– salivary glands s. parotid
  gland, s. submandibular
  gland
– scalp 9
– seminal vesicles 117–119
– skin, cheek s. face
– – neck 9
– – penis 120
– – scrotum 120

– small-bowel 88
– spleen 85
– stomach *Fig. 40*, 84, 85
– submandibular gland 15
– suprarenal gland 90
– testicle *Fig. 55*, 114, 118,
  119
– thymus gland 46
– thyroid gland *Fig. 8*, 15
– tongue 12
– tonsils s. oropharynx
– ureter 91
– urethra 117, 120
– urinary bladder *Fig. 53*,
  114, 115, 117, 118, 126
– uterus, cervix 114, 123, 124,
  127
– – body *Fig. 56*, 114, 122
– vagina 114, 123, 124, 126
– – vestibulum *Fig. 56*. 124
– vulva, labia pudendi
  Fig. 56, 124
lymphatic duct *Figs. 3, 4*
– right lymphatic duct
  Fig. 21, 3–6, 41, 44, 46, 48
– thoracic duct *Figs. 21, 24,
  38*, 1, 4–6, 41, 4446, 48, 82
lymphatic trunk *Figs. 3, 4*
– bronchomediastinal 4, 6,
  41, 46
– intestinal *Fig. 38*, 4, 6, 82,
  88, 93
– jugular 3, 5, 7, 41, 46
– lumbar *Fig. 38*, 3, 6, 82, 88,
  91, 125
– subclavian 3, 5, 41, 46, 49
lymph nodes
– anterior cervical s. cervical
– anterior mediastinal s. mediastinal
– apical s. deep axillary
– appendicular 88, 89, 94
– axillary *Figs. 23, 24*, 48, 51
– – deep *Figs. 26, 27*, 48, 51
– – – apical *Fig. 25*, 49, 51

– – – central *Figs. 25, 26,*
49, 51
– – interpectoral *Figs. 26, 29,*
30, 48, 49, 51
– – superficial *Figs. 27–30,*
48, 49, 51
– – – lateral 49, 51
– – – pectoral 48, 51
– – – subscapular 48, 51, 82
– bronchopulmonary s. posterior mediastinal
– buccinator node s. facial
– cecal, pre- and retrocecal
Fig. 57, 88, 89, 94
– celiac *Figs. 38, 39, 41,* 47,
53, 83, 85, 87, 88, 90, 93
– central s. deep axillary
– cervical, anterior *Figs. 5,*
15–18, 25, 13, 14, 15, 18
– – infrahyoid 14, 18
– – paratracheal *Figs. 5, 17,*
18, 13–15
– – prelaryngeal *Figs. 5, 8,*
16, 13–15, 18
– – pretracheal *Figs. 5, 17,*
18, 14, 15, 18
– – superficial *Figs. 5, 16,*
18
– – suprasternal *Fig. 18,* 18
– – thyroid *Figs. 5, 18,*
15, 18
– cervical, lateral *Figs. 5,*
11–18, 25
– – deep *Figs. 5, 8, 13–18,*
9–15, 18, 44, 47, 51
– – – jugular, anterior and
lateral *Figs. 5, 12, 18,*
19, 46, 47, 51
– – – jugulodigastric *Fig. 5,*
7, 13, 19
– – – jugulo-omohyoid
Fig. 5, 19
– – – retropharyngeal
Figs. 5, 6, 11–18,
10–13, 19, 46, 47, 51
– – – supraclavicular
Figs. 5, 17, 18, 23,
9, 14, 19, 49
– – superficial *Figs. 5, 14, 15,*
17, 18, 9, 18
– colic s. mesocolic
– common iliac s. iliac
– cystic node s. hepatic
– deep s. axillary, lateral,
cervical and inguinal

– epigastric, inferior 82, 92
– facial *Figs. 5, 10,* 17
– – buccinator node *Fig. 11,*
17
– – malar node 17
– – mandibular node
Fig. 12, 17
– – nasolabial node 17
– foraminal node s. hepatic
– gastric, right and left
Figs. 22, 41, 43, 44, 82, 85,
93
– gastroomental *Figs. 43–45,*
85, 93
– gluteal, superior and
inferior s. internal iliac
– hepatic *Figs. 39, 41, 45,*
82–86, 88, 92, 93
– – cystic node *Fig. 45,*
84, 93
– – foraminal node 84, 93
– ileocolic *Fig. 38,* 88, 89,
94
– iliac *Figs. 38, 51, 53,* 114
– – common iliac
Figs. 54, 56–58,
91, 114, 118, 123–126
– – – intermediate 125
– – – lateral 115, 125
– – – medial 125
– – – promontory *Fig. 58,*
115, 117, 118, 123–125
– – – subaortic 122, 123,
125
– – external iliac
Figs. 54–56, 59–61, 64, 65
91, 114, 116, 118–120,
122–127
– – – interiliac 126
– – – intermediate
116–120, 122–124, 126
– – – lacunar
– – – – intermediate 126
– – – – lateral 118, 120,
125, 126
– – – – medial 116, 117,
120, 124–126
– – – lateral 116, 126
– – – medial 116–119, 123,
124, 126
– – – obturator *Fig. 54,*
124, 126
– – internal iliac
Figs, 54, 56, 59, 64,
82, 114–119, 122–125

– – – gluteal, superior und
inferior *Figs. 60, 64,*
117, 118, 122, 124, 126
– – – sacral *Figs. 54, 59, 60,*
123, 126
– inferior mesenteric s.
mesenteric
– inferior phrenic s. phrenic
– infrahyoid s. anterior
cervical
– inguinal *Fig. 51,*
82, 114, 126
– – deep *Figs. 56, 62, 63, 66,*
120, 124, 125
– – superficial
Figs. 56, 62, 63, 66,
114, 115, 117, 120, 122,
124, 125, 127
– – – inferior 127
– – – superolateral 127
– – – superomedial 122, 127
– intercostal *Figs. 21, 24,*
26–37, 43, 48, 52, 82
– interiliac s. external iliac
– intermediate s. common
iliac, external iliac and
lumbar
– interpectoral s. axillary
– jugular anterior and lateral
s. lateral cervical
– jugulodigastric s. lateral
cervical
– juguloomohyoid s. lateral
cervical
– juxta-esophageal pulmonary
s. posterior mediastinal
– lacunar, intermediate, lateral
and medial s. external iliac
– lateral s. external iliac,
common iliac and
superficial axillary
– lateral cervical s. cervical
– lateral vesical s. paravesicular
– lateroaortic s. left lumbar
– laterocaval s. right lumbar
– left lumbar s. lumbar
– ligamentum arteriosum
node s. mediastinal
– lumbar *Figs. 38, 51, 55,*
82–85, 90, 114, 119, 122,
125
– – left lumbar
Figs. 46–50, 54,
87, 90, 91, 119, 120, 122

– – – lateroaortic *Fig. 56,*
   82, 84, 87, 88, 90, 91,
   95,114,120, 122
– – – preaortic *Fig. 56,* 84,
   88, 90, 91, 95, 114, 120,
   122
– – – postaortic 91
– – intermediate lumbar
   Figs. 47–49, 56, 84, 86,
   87, 90–92,120
– – right lumbar *Figs. 45–50,*
   91, 92, 119, 120, 122
– – – laterocaval
   Figs. 46, 56, 92, 120
– – – precaval *Figs. 46, 56,*
   90, 92, 120
– – – postcaval *Figs, 46–48,*
   58, 92
– malar node s. facial
– mandibular node s. facial
   mastoidal *Figs. 5, 10–12,*
   9, 11, 17
– medial s. common iliac,
   external iliac
– mediastinal
– – anterior mediastinal
   Figs. 20, 25–34,
   43, 44, 46, 48, 52, 82
– – node at the ligamentum
   arteriosum *Fig. 31,*
   43, 52
– – posterior mediastinal
   Figs. 20, 21, 39,
   43, 44, 48, 53, 82, 83, 85,90
– – – bronchopulmonary
   Figs. 19, 32–35,
   42, 43, 53
– – – juxta-esophageal
   pulmonary
   Figs. 30–36,
   42, 44, 45, 53
– – – node at the arch of the
   azygos vein *Fig. 31,*
   43, 44, 53
– – – paratracheal
   Figs. 19, 20, 25–30,
   16, 43, 44, 46, 47, 53
– – – tracheobronchial,
   superior and inferior
   Figs. 19, 20, 30–33,
   42–44, 46, 53
– – – subcorynal
   Figs. 19, 20, 44
– mesenteric *Figs. 38, 41, 42,*
   82, 83, 88, 89, 93, 94

– – inferior 88, 89, 94, 114
– – – sigmoidal *Fig. 58,*
   88, 89, 94, 114
– – – superior rectal 94,
   114, 115, 125
– – juxtaintestinal
   Figs. 45–50, 57, 58, 64, 94
– – superior *Figs. 46–49,*
   86–89, 94
– mesocolic *Figs. 38, 42, 44,*
   48–50, 89, 94
– – colic, right, medial
   and left *Fig. 57,* 89, 94
– – paracolic *Figs. 46, 47, 49,*
   57, 58, 88, 89
– nasolabial node s. facial
– nuchal *Figs. 13–16,* 7, 19
– obturator s. external iliac
– occipital *Figs. 5, 10–12,* 17
– pancreatic, superior and
   inferior *Figs. 41, 45–47,*
   87, 93
– pancreaticoduodenal,
   superior and inferior
   Figs. 41, 46–48,
   84, 86–88, 93
– paracolic s. mesocolic
– paramammary
   Figs. 32–36, 49, 51, 52, 82
– pararectal *Figs. 59–62,*
   64–66, 113–115, 124, 127
– parasternal
   Figs. 23, 24, 29–36, 39,
   44, 48, 51, 52, 82, 83
– paratracheal s. posterior
   mediastinal and anterior
   cervical
– parauterine *Figs. 56, 64, 65,*
   113, 122–124, 127
– paravaginal *Fig. 66,* 113,
   123, 124, 127
– paravesicular *Fig. 53,* 113,
   116, 118, 126
– – lateral vesical *Figs. 60,*
   65, 126
– – postvesicular *Figs. 60,*
   65, 126
– – prevesicular *Figs. 51, 61,*
   117, 126
– parotid, deep and superficial
   Figs. 5, 10–13, 9, 11, 15, 17
– pectoral s. superficial
   axillary pericardial, lateral
   Figs. 36, 37, 39, 46,
   52, 83

– phrenic, superior
   Figs. 37, 43–46, 48, 53, 83
– – inferior *Figs. 43, 44,* 92
– postaortic s. left lumbar
– postcaval s. right lumbar
– posterior mediastinal
   s. mediastinal
– postvesicular s. paravesicular
– preaortic s. left lumbar
– preauricular s. parotid
– precaval s. right lumbar
– prelaryngeal s. anterior
   cervical
– prepericardial *Figs. 35–37,*
   46, 52
– pretracheal s. anterior
   cervical
– prevertebral
   Figs. 25–30, 33, 34, 36, 37, 44,
   48, 53
– prevesicular s. paravesicular
– promontory s. common iliac
– pyloric *Fig. 41,* 85, 88, 93
– – retropyloric 93
– – subpyloric 93
– – suprapyloric 93
– retropharyngeal s. lateral
   cervical
– retropyloric s. pyloric
– right lumbar s. lumbar
– Rotter's nodes s. interpectoral axillary
– sacral s. internal iliac
– sentinel nodes s. deep cervical
– sigmoidal s. inferior
   mesenteric splenic
   Figs. 44, 45, 85, 87, 93
– subaortic s. common iliac
– subcorynal s. posterior
   mediastinal
– sublingual *Fig. 12,* 12, 18
– submandibular *Figs. 5,*
   12–15, 9–13, 15, 17
– submental *Figs. 5, 13–15,*
   9, 12, 18
– subpyloric s. pyloric
– subscapular s. superficial
   axillary
– superficial s. anterior
   cervical
– – lateral cervical
– – axillary and inguinal
– superior phrenic s. phrenic

– superior rectal s. inferior mesenteric
– supraclavicular s. anterior cervical
– suprapyloric s. pyloric
– suprasternal s. anterior cervical
– thyroid s. anterior cervical
– tracheobronchial s. posterior mediastinal
lymphography 1
lymphoszintigraphy 1

right lymphatic duct s. lymphatic duct

sonography of lymph nodes 1
subclavian trunk s. lymphatic trunk

thoracic duct s. lymphatic duct

Printing and Binding: Strauss GmbH, Mörlenbach